Changing Syntheses
in
Development

Changing Syntheses in Development
The Twenty-Ninth Symposium
The Society for Developmental Biology

Albany, New York, June 17–19, 1970

EXECUTIVE COMMITTEE
1969–1970

MEREDITH N. RUNNER, University of Colorado, *President*
ANTON LANG, Michigan State University, *Past-President*
FOLKE SKOOG, University of Wisconsin, *President-Designate*
PAUL B. GREEN, University of Pennsylvania, *Secretary*
ELIZABETH D. HAY, Harvard Medical School, *Treasurer*
DONALD D. BROWN, Carnegie Institution of Washington
MAC V. EDDS, JR. Brown University, *Editor-in-Chief*

1970–1971

FOLKE SKOOG, University of Wisconsin, *President*
MEREDITH N. RUNNER, University of Colorado, *Past-President*
FRANK H. RUDDLE, Yale University, *President-Designate*
MALCOLM S. STEINBERG, Princeton University, *Secretary*
ELIZABETH D. HAY, Harvard University, *Treasurer*
PAUL B. GREEN, University of Pennsylvania
MAC V. EDDS, JR. Brown University, *Editor-in-Chief*

Business Manager

W. SUE BADMAN
P.O. Box 2782
Kalamazoo, Michigan 49003
616-345-7476

Changing Syntheses in Development

Edited by

Meredith N. Runner

Molecular, Cellular, and Developmental Biology
University of Colorado
Boulder, Colorado 80302

Developmental Biology, Supplement 4

Editor-in-Chief

M. V. Edds, Jr.

1970

ACADEMIC PRESS, New York and London

Copyright © 1971, by Academic Press, Inc.
ALL RIGHTS RESERVED.
NO PART OF THIS BOOK MAY BE REPRODUCED IN ANY FORM, BY PHOTOSTAT, MICROFILM, BY RETRIEVAL SYSTEM, OR ANY OTHER MEANS, WITHOUT WRITTEN PERMISSION FROM THE PUBLISHERS.

ACADEMIC PRESS, INC.
111 Fifth Avenue, New York, New York 10003

United Kingdom Edition published by
ACADEMIC PRESS, INC. (LONDON) LTD.
Berkeley Square House, London W.1

LIBRARY OF CONGRESS CATALOG CARD NUMBER: 55-10678

PRINTED IN THE UNITED STATES OF AMERICA

Changing Syntheses in Development

The 29th Symposium was held at Albany, New York, June 17-19, 1970. The Society gratefully acknowledges the efficiency of the host committee, the hospitality of the State University of New York and the support from the National Science Foundation.

Contributors and Presiding Chairmen

Numbers in parentheses indicate the pages on which the authors' contributions begin

I. Changing Syntheses

Chairmen. Philip Grant, Department of Biology, University of Oregon, Eugene, Oregon 97403, and Melvin Spiegel, Department of Biology, Dartmouth, Hanover, New Hampshire 03755.

RICHARD C. STARR, Department of Botany, Indiana University, Bloomington, Indiana 47401 (59)

THOMAS H. SHEPARD AND TAKASHI TANIMURA, Department of Pediatrics, AND MAURICE A. ROBKIN, Department of Nuclear Engineering, University of Washington, Seattle, Washington 98105 (42)

R. A. FLICKINGER, Department of Biology, State University of New York, Buffalo, New York 14214 (12)

Discussion. Localization of developmental information in the egg and early embryo. Martin Nemer, Institute for Cancer Research, Philadelphia, Pennsylvania 19111, Denis Smith, Biology Department, Purdue University, Lafayette, Indiana 47907

Discussion. Synthesis of macromolecules during growth and differentiation. Tom Humphreys, Department of Biology, University of California, San Diego, California 92037, Joram Piatigorsky, National Institute of Child Health and Human Development, National Institutes of Health, Bethesda, Maryland 20014

II. Controls for Special Syntheses

Chairmen. John R. Coleman, Division of Biological and Medical Sciences, Brown University, Providence, Rhode Island 02912 and Irwin Konigsberg, Department of Biology, University of Virginia, Charlottesville, Virginia 29903

YALE J. TOPPER, S. H. FRIEDBERG, AND TAKAMI OKA, National Institute of Arthritis and Metabolic Diseases, National Institutes of Health, Bethesda, Maryland 20014 (101)

MARTIN POSNER, WILLIAM J. GARTLAND, JEFFREY L. CLARK, AND GORDON SATO, Department of Biological Sciences, University of California, San Diego, California 92037, AND CARL A. HIRSCH, Department of Medicine, Beth Israel Hospital, Boston, Massachusetts 02215 (114)

ALBERT DORFMAN, Department of Pediatrics, University of Chicago, Chicago, Illinois 60637

Discussion. Cell commitment and phenotypic expression. Stuart Haywood, Department of Biological Sciences, University of Connecticut, Storrs, Connecticut 06268. Jay Lash, Department of Anatomy, University of Pennsylvania, Philadelphia, Pennsylvania 19104

Discussion. Regulation of synthesis in relation to the cell cycle. Robert R. Klevecz, Division of Biology, City of Hope Medical Center, Duarte, California 91010, David W. Martin, Department of Biochemistry, University of California Medical Center, San Francisco, California 94122

III. The Role of Collagen

Chairmen. Elizabeth D. Hay, Department of Anatomy, Harvard Medical School, Boston, Massachusetts 02115, and Edgar Zwilling, Department of Biology, Brandeis University, Waltham, Massachusetts 02154

ZACHARIAS DISCHE, Department of Ophthalmology, College of Physicians and Surgeons, Columbia University, New York, New York 10032 (164)

MARSHALL R. URIST, Bone Research Laboratory, UCLA School of Medicine, Los Angeles, California 90024 (125)

MERTON R. BERNFIELD, Department of Pediatrics, AND NORMAN K. WESSELS, Department of Biological Sciences, Stanford University, Stanford, California 94305 (195)

Discussion. Origin and morphogenetic functions of collagen. Alfred J. Coulombre, National Eye Institute, National Institutes of Health 20014, Jerome Gross, Developmental Biological Laboratory, Massachusetts General Hospital, Boston, Massachusetts 02114

Discussion. Significance of early appearance of differentiated products. Howard Holtzer, Department of Anatomy, University of Pennsylvania, Philadelphia, Pennsylvania 19104

Chairman of Host Committee: Joseph Mascarenhas, Department of Biological Sciences, State University of New York at Albany, Albany, New York 12203

Contents

CONTRIBUTORS AND PRESIDING CHAIRMEN v

I. Changing Syntheses

Changing Syntheses in Development
   MEREDITH N. RUNNER

      Introduction 1
      Analytic, Pragmatic Approaches 2
      Cellular and Transcriptional Correlates—Phenomenology 5
      Summary 10
      References 11

The Role of Gene Redundancy and Number of Cell Divisions in Embryonic Determination
   R. A. FLICKINGER

      Introduction 12
      Relation between Rate and Number of Cell Divisions and RNA Synthesis 12
      Relation of Cell Division to a Particular Pathway of Determination 25
      Possible Control Mechanisms for Transcription and Determination 27
      Evolutionary Implications of the Role of Gene Redundancy in Cellular Determination 34
      Summary 38
      References 39

Energy Metabolism in Early Mammalian Embryos
   THOMAS H. SHEPARD, TAKASHI TANIMURA, AND
   MAURICE A. ROBKIN

      Introduction 42
      Biological Material 42
      Whole Embryo *in Vitro* Studies 43
      Heart Rate Changes at Different Oxygen Concentrations 46
      Glucose Metabolism 47

Terminal Electron Transport System	52
Intercorrelations of Oxygen Requirement, Glucose Metabolism, and the Terminal Electron Transport System	54
Summary	56
References	57

Control of Differentiation in Volvox

RICHARD C. STARR

Introduction	59
Volvox carteri f. *nagariensis* Iyengar	61
Embryogenesis	61
Release of the Young Spheroids	73
Sexual Fusion and Zygote Germination	73
The Inducer	74
Initiation of the Sexual Response	79
Mutants Affecting the Developmental Pattern	81
The Determination of Somatic Cells	84
Volvox aureus Ehrenberg	89
Induction of Males	89
Induction of Parthenospores	92
Differentiation in the Embryo	93
Volvox rousseletii f. *griquaensis* Pocock	93
Induction of Male and Female Spheroids	93
The Inducer	95
Volvox gigas Pocock	95
Other Species of *Volvox*	96
Summary	97
References	99

II. Controls for Special Syntheses

On the Development of Insulin Sensitivity by Mouse *Mammary Gland in Vitro*

YALE J. TOPPER, S. H. FRIEDBERG, AND TAKAMI OKA

Introduction	101
Materials and Methods	102
Results	102
Discussion	110
References	112

Development of Hormone-Responsive Cell Strains *in Vitro*

MARTIN POSNER, WILLIAM J. GARTLAND, JEFFREY L. CLARK, GORDON SATO, AND CARL A. HIRSCH

Introduction	114
Selection for Hormone-Dependent Cell Strains	115

Development of a Hormone-Dependent Mammary
  Tumor Cell Line 116
Experiments Using Hormone-Dependent Cell Strains . 117
Summary 123
References 123

III. The Role of Collagen

The Substratum for Bone Morphogenesis

MARSHALL R. URIST

Introduction 125
Materials and Methods 127
Results 128
  Cell Differentiation and Bone Morphogenesis in
    Implants of Bone Matrix 128
  Ultrastructure of the Substratum and Cell Interfaces
    before, during, and after Bone Morphogenesis . . 133
  Points of Loose Contact Between Osteoprogenitor Cells
    and Old Bone Matrix 141
  Sequence of Cell-Substratum Interfaces with Cell Differentiation 144
  Yield of New Bone from Normal Matrix 145
  Preparations of Cold HCl-Demineralized Matrix which
    do not Impose Bone Morphogenetic Pattern upon
    Mesenchyme 146
Discussion 149
  Kinetics of Cell-Substratum Interactions 150
  Ultrastructure of the Bone Morphogenetic Interface 151
  Chemical Morphology of the Bone Morphogenetic
    Matrix 153
  Hypothetical Coupling of Collagen Cross-Linkages and
    the Calcification Initiator to Produce the Bone
    Morphogenetic Pattern 159
Summary 159
References 161

Collagen of Embryonic Type in the Vertebrate Eye and Its Relation to Carbohydrates and Subunit Structure of Tropocollagen

ZACHARIAS DISCHE

Introduction to Morphology of Collagen and Its Possible
  Relation to Carbohydrates 164
Nonfibrous Collagen of the Lens Capsule 171
Corneal Stroma: Connective Tissue Arrested at an Embryonal Stage of Development 178
Biological Implications 189
Summary 190
References 191

Intra- and Extracellular Control of Epithelial Morphogenesis
MERTON R. BERNFIELD AND NORMAN K. WESSELLS

    Introduction 195
    Extracellular Control of Morphogenesis 196
    Acid Mucopolysaccharide at the Epitheliomesenchymal
      Junction 203
    Surface-Associated Mucopolysaccharide and Salivary
      Gland Morphogenesis 207
    Characteristics of the Surface-Associated Material . . 220
    Intracellular Control of Morphogenesis 223
    Microfilaments and Salivary Gland Morphogenesis . . 227
    Concluding Remarks 239
    References 244
AUTHOR INDEX 250
GLOSSARY OF ABBREVIATIONS 255
SUBJECT INDEX 256

29th SYMPOSIUM, JUNE 17–19, 1970
STATE UNIVERSITY OF NEW YORK AT ALBANY

Opening remarks—Robert D. Allen

W. Sue Badman   Elizabeth D. Hay   Folke K. Skoog
       Anton Lang   Herbert Stern

Joseph Mascarenhas   Donald L. Kimmel
     Joan Abbott   John R. Coleman

James H. Gregg   A. C. Clement   Laurens N. Ruben
   Melvin Spiegel   Charles G. Melton   Alex J. Haggis

Meredith Runner   Eloise E. Clark   Robert D. Allen

Erwin R. Konigsberg   Salome Gluecksohn-Waelsch
          Alfred J. Coulombre

Intermission—Lecture Center

Quentin N. LaHam   Jack Greenberg   Reed Flickinger

Stuart Haywood   Richard Landesman   Arnold I. Caplan

Changing Syntheses in Development[1]

MEREDITH N. RUNNER

*Department of Molecular, Cellular, and Developmental Biology,
University of Colorado, Boulder, Colorado 80305*

INTRODUCTION

"Every step of development is a physiological reaction, involving a long and complex chain of cause and effect between the stimulus and the response. The character of the response is determined not by the stimulus, but by the organization." [page 430] "It is the remarkable substance nuclein*,—which is almost certainly identical with chromatin,—that chiefly claims attention here on account of the physiological role of the nucleus." [page 332] "If chromatin inheres the sum total of hereditary forces, and if it be equally distributed at every cell division, how can its mode of action vary in different cells to cause diversity of structure, e.g. differentiation?" [page 413]

* Nuclein is degradable into protein, phosphoric acid, a carbohydrate and nuclein bases such as adenine and guanosine. [page 336]

(Wilson, 1902).

This symposium of the Society for Developmental Biology, like the twenty-eight symposia which went before it, attempts to serve both a pedagogical and an investigational role. The published product of this symposium, a collection of reviews depicting changing syntheses during several phases of cell and tissue interactions, should be of interest to graduate students and teachers. The theme running throughout the volume is emphasis upon changes which bridge the gap between the biochemistry and the sculpturing of a new organism. This assembly of reviews on changing syntheses during cell and tissue interactions, perhaps a departure from the current trend toward molecular reductionalism, is intended to cover more than biochemical differentiation; perhaps it is an overambitious intention that the collection of reports will relate cell and tissue interactions to molecular biology, ultrastructure, and morphogenesis.

Irrespective of whether developmental biologists favor a concept of environmental or extracellular evocators interacting with a system

[1] The 29th Symposium was held at Albany, New York, June 17-19, 1970. The Society gratefully acknowledges the efficiency of the host committee, the hospitality of the State University of New York and the support from the National Science Foundation.

of gene regulator substances (Waddington, 1967) or whether they favor activation of a system of gene slaves (Callan, 1967; Britten and Davidson, 1970) in order to account for epigenetic changes during development, RNA is the common link mediating the program of nuclear chromatin with prospective products or capabilities characterizing cellular phenotypes. It is this common link, DNA-like-RNA (D-RNA), which this chapter attempts to correlate with changing syntheses.

ANALYTIC, PRAGMATIC APPROACHES

The volume attempts to show that investigators have taken advantage of two pragmatic approaches: (a) pursuit of the aftermaths of known events, such as fertilization, transplantation, or metabolic inhibitors, which impose environmentally induced changes in the developing system; and (b) detection of the onset of new phenotypes, products, capabilities, or other expressions. These two pragmatic approaches, the aftermaths of known events and the detection of new products, permit functional correlates to be made between changing synthesis and ontogenetic phenomena, i.e., the major achievements of changing synthesis in development.

Functional correlates between cellular and molecular approaches to development can be grouped into four ontogenetic phenomena (see Fig. 1): (a) rapid multiplication, (b) cellular commitments, (c) initiation of expressions, and (d) modulation of existing expressions. Although depicted sequentially in Fig. 1, these correlates are not mutually exclusive and represent, at best, gradual transitions. (a) *Exponential growth* in the early embryo is associated with specialization of syntheses of precursors and precursor enzymes for mitosis and for polymerization of nucleic acids. (b) Cells become *committed* or determined as RNA copies of genes shift from the highly repetitive to the less repetitive populations.[2] For this phenomenon a model of D-RNA epigenesis is proposed. (c) *Initiation* of cellular products or expressions along with increase in effectiveness of existing gene copies is postulated for the epigenesis of cell products as illustrated in this volume by Bernfield and Wessells. (d) *Regulation* of existing specialized cellular expressions tends to be associated with peak activities of copies from nonrepetitive DNA sequences.

[2] Speculations such as this one are direct outgrowths of contributions by authors of the succeeding chapters. In no way, however, can the authors be implicated in the fallacies inherent in such speculations.

Cellular Phenomena	Quantitative and Qualitative Changes in D-RNA
A. RAPID MULTIPLICATION Prospective potency exceeds ultimate fate Exponential growth	A. DOMINANCE OF MULTIPLE GENE COPIES Evolutionarily old and repetitive DNA sequences for sustenance and cell division *Specialized syntheses for nucleic acids* Energy utilized for high proportions of RNA per cell Nucleotide pathways
B. EMERGENCE OF CELLULAR COMMITMENTS Determination, becomes "set" Regionalization can be demonstrated Potency becomes restricted (Determination can be delayed) (Competence can be prolonged) Cellular and tissue interactions Predifferentiation	B. INCREASE IN COPIES OF LESS REPETITIVE DNA SEQUENCES Recapitulates phylogeny *Slower division rates* Uncouples DNA and RNA synthesis Increases relative times for S or G_1 Favors less redundant gene copies Nuclear RNA changed qualitatively by *absolute number of cell divisions* Older and more redundant gene copies are attenuated by *late replication*
C. INITIATION OF NEW CELLULAR EXPRESSIONS Protodifferentiation Luxury products for export Prematurely inducible (cortisone) Indefinitely suppressible (BUdR) Muscle fusion	C. INCREASE EFFECTIVENESS OF EXISTING GENE COPIES *Transport of D-RNA from nucleus to cytoplasm* *Selective conservation of D-RNA in cytoplasm* Association of RNA with ribosomes and membranes
D. REGULATION OF CELLULAR EXPRESSION Reversible dedifferentiation Resumption of expression Modulation	D. DOMINANCE OF RNA COPIES FROM NONREPETITIVE DNA Terminal differentiation Permanent messengers

FIG. 1. Changing syntheses—a model of D-RNA epigenesis [*See footnote page 2*]. Cellular and transcriptional correlates from cleavage to organogenesis.

The following comments are intended to superimpose such a plan of organization upon the articles in the volume and to correlate cellular phenomena with quantitative and qualitative changes in D-RNA and with cellular responsiveness to environmental and experimental manipulation.

Detection of New Products and Capabilities in Ontogeny

Detection of smaller and smaller quantities of products and smaller and smaller biochemical differences during ontogeny provides experimental end points in search of initiating events. Macromolecules for export, new enzymes, or new or unique species of RNA illustrate types of products to be investigated in order to understand mechanisms for initiation of expression, or, as often put, control mechanisms. Discovery of new biological systems with clean end points to confirm fashionable dogma seems to be highly rewarding. Flickinger's experiments seem particularly refreshing because they contribute framework for a model for sequential syntheses in the embryo.

Sequelae Following Well-Defined Events

Abrupt changes in ontogeny, such as fertilization, cellular interactions, hormonal stimulation, or consequences of metabolic inhibitors, provide triggering mechanisms from which sequential changes can be investigated.

Topper, Friedberg, and Oka describe the consequences of an event —removal of virgin mammary gland epithelium to culture—which initiates intracellular changes in syntheses. Induction of change in synthesis in the explanted virgin mammary epithelium requires 24–48 hours, depending upon the technique to make the specific measurements and the parameter used to measure the lag period.

The sequelae approach attempts to exploit ontogenetic events during changing syntheses. A key question is whether induced biochemical changes observed [e.g., change in D-RNA populations (Flickinger) or reduced electron transport mechanism (Shepard *et al.*)] are incidental or concomitant changes or whether they are in the causal chain of events leading to morphogenesis. Changing syntheses in development therefore include searches for controlling mechanisms starting with detectable onset of new products or starting with detectable sequences of changes subsequent to abrupt morphogenetic events.

CELLULAR AND TRANSCRIPTIONAL CORRELATES—PHENOMENOLOGY

Exponential Growth and Redundant Gene Copies

Rapidly succeeding cell divisions are evolutionarily associated with dominance of copies of old and repetitive genes; those necessary for cell division and sustenance. The volvox embryo described by Starr undergoes twelve rapid cleavages in the absence of appreciable G_1, or growth, phases. The end result is approximately four thousand cells which presumably have a full complement of DNA (haploid). This illustrates embryonic specialization for exponential synthesis of precursor enzymes and polymerization of their products into nucleic acids. Shepard reports that the early somite stages of rat embryos utilized glucose at a rate about ten times that found in adult liver tissue. Investigations of energy balance sheets required for exponential synthesis of bases and nucleotides during this specialized period for nucleic acid synthesis are indicated. High energy utilization and specialization for nucleic acid synthesis correlate well with requirements for evolutionarily old and redundant genes so necessary for sustenance and cell division, i.e., exponential growth.

Onset of Cellular Commitment and Increase in Copies of D-RNA in Relation to Homologous DNA Sequences.

Early determinations occur during exponential growth and are detected by subsequent cellular behavior in isolation. How cells become committed or determined is a major problem. Starr shows that an inducer substance is released into the medium as volvox releases sperm. This inducer substance can be assayed in a female strain where newly liberated asexual spheroids are transformed into sexual females. The inducer substance commits the gonidia to become egg initials at an effective concentration of about $3 \times 10^{-14} M$. Under the influence of this powerfully effective inducer substance, what would have been asexual embryos with 16 gonidia at the fifth cleavage now become determined to be sexual females, i.e., to differentiate two cleavages later and to possess 40 egg initials.

Flickinger's experiments address themselves to changing populations of gene copies as development progresses and extend the work of Denis (1968). Flickinger's estimations for the onset of appearance of new kinds of D-RNA are, in fact, attempts to determine when and how commitments of cells have been initiated. His report stresses cellular commitment as phylogeny tends to recapitulate itself and multiple gene copies phase out while less redundant gene copies become more effective. His studies on differential rates of division between gastrulation and swimming larvae attempt to uncouple DNA and RNA synthesis and thereby permit detection of precocious RNA synthesis. Slowing the division rate, either experimentally or as occurs naturally during ontogeny, increases the relative duration of the S or G_1 phases, thereby favoring the less redundant gene copies. Flickinger's experiments study both the absolute number of cell divisions and the rates of cell divisions. The experiments for absolute number of cell divisions show that the D-RNA's became qualitatively changed as the number of divisions progressed. Chromosomal studies in Flickinger's laboratory report that specific, late replications occur in different regions of the embryo. He concludes that late replications reduce opportunity for production of redundant gene copies. Thereby, at the onset of cellular commitment the balance shifts from more to less redundant gene copies.

Initiation of New Expressions as Effectiveness of Gene Copies Increases

Products or capabilities, not previously detected, become detectable during ontogeny. Therefore the embryonic machinery is either fully programmed, and no initiating event is required, or some extracellular event initiates changing syntheses. After having become fully capable and committed to change synthesis, some extracellular stimulus seems necessary to initiate new expressions. The interaction of cells and tissues and the appearance of fine structural organelles, as presented by Bernstein and Wessells, demonstrate initiation of ultrastructural and biochemical changes associated with new expressions.

Initiation of capability (response to insulin) has been followed by Topper *et al.* in epithelial cells of the explanted mammary gland of adult, virgin mice. After removal of the virgin glands to explant, these nonproliferative, virgin glands acquire the capability (response to insulin) found in mammary epithelium from pregnant mice. (Beware of cell culturists who suggest that behavior of cells in explant may reflect capabilities existing *in vivo* at the moment of explant.) These nonproliferative glands, during the first few hours after explant, fail to respond to insulin. The capabilities of changing syntheses to respond to insulin are measured by uptake of α-aminoisobutyric acid into protein and uptake of thymidine into DNA. They are also measured as enzymatic activities of cytoplasmic glucose-6-

phosphate dehydrogenase and as membrane-associated enzyme, NADH-cytochrome *c* reductase. Generalized cellular responses to insulin by cytoplasmic and membranous enzymes and protein and DNA synthesis begin within less than 24 hours. This suggests that changing syntheses may be due to activation of existing machinery of the cell, such as rough endoplasmic reticulum, to effectively use existing precursors with which to initiate the new capabilities.

Bernfield and Wessells address themselves to analysis of cellular and tissue interactions (induction) as they lead to cytodifferentiation (appearance of unique cell types) and to morphogenesis (appearance of unique populations of cells). Their biological system for studying determination, cytodifferentiation, morphogenesis, and initiation of new expressions is the epitheliomesenchyme interaction in the salivary gland. Previously it had been shown that epithelial cells stimulate mesenchymal cells to produce precursor to collagen and that collagen polymerizes to form a portion of the basement membrane of the epithelial cells. Careful biochemical dissection of the basement membrane has revealed several interfacial materials which participate in morphogenesis. The components of the basement membrane are collagen, large and small glycoproteins, neutral polysaccharides, and acid mucopolysaccharide. Bernfield and Wessells supply evidence to indicate that acid mucopolysaccharide is a surface active, morphogenetic agent at the epitheliomesenchymal junction.

Dische reports that a number of factors may influence the initial formation and the higher organization of fibrils from tropocollagen. Substances such as chondroitin sulfate, galactoglucan, sialoglycohexoaminoglycans, and acid mucopolysaccharide can modify collagen by adhering to the tropocollagen molecules. Dische draws a structural analogy from the corneal stroma when he proposes that during embryogenesis a three-dimensional network of tropocollagen interacts with fibrillar collagen. This interaction is mediated by polysaccharide to produce three-dimensional ionic forces which may play a role in the morphogenesis of epithelial cells interacting with mesenchyme. Dische further hypothesizes that carbohydrate-rich subunits of collagen may be carriers for proteins which participate in the process of embryonic induction. Investigators seem to agree that acid mucopolysaccharides produced by epithelial cells influence the arrangement of fibrillar collagen with a consequent effect on morphogenesis at the epithelial-mesenchymal interface. It becomes increasingly apparent that regulated mixtures of new products in extracellular spaces may participate in changing syntheses resulting in morphogenesis and initiation of new phenotypic characteristics. Thus findings of Dische support the proposal by Bernfield and Wessells that acid mucopolysaccharide is the active, morphogenetic agent.

A dynamic model for the morphological branching pattern seen in the submaxillary, salivary gland of the mouse invokes epithelial cell shape, epithelial cell multiplication, and integration of intracellular and exported products (Bernfield and Wessells). A model for initiating epithelial morphology is based upon extracellular deposition of a mucopolysaccharide–protein complex, which by itself has a cell-flattening tendency, interacting with adjacent intracellular microfilaments, which by themselves produce a cell-rounding tendency. Localized (linear?) polymerization of mucopolysaccharides results in contraction of adjacent microfilaments. This regional (linear?) contraction tends to produce a purse-string effect on expanding epithelium with a resulting cleft between areas of continuing expansion and evagination. Collagen becomes polymerized inside the cleft, thereby stabilizing the new morphology. At the same time the outermost portion of the evaginations acquires a new layer of mucopolysaccharide from the epithelial cells, stimulates contraction of the fine filaments, produces a subsequent cleft which then becomes stabilized by condensation of collagen. The model of bulges and clefts stabilized by collagen results in the branching pattern seen in epithelial glands. The coordination between the extracellular deposition of mucopolysaccharides and the corresponding intracellular bundles of microfilaments illustrates initiation of new cellular expressions possibly derived from increased effectiveness of existing copies of slightly redundant genes.

Maintenance and Control of Previously Established Expression. Dominance of Unique Gene Copies

Once an expression has been initiated, certain controls are exerted which can reduce, shut down, turn on, or sustain previously determined expressions. Modulation controls can be seen in the oscillation of syntheses within the cell cycle and hormonal controls of certain special syntheses. Induced changes in mammary glands described by Topper *et al.* show that a number of parameters are changed (regulated) by insulin once tissue sensitivity is acquired. Isolation of special expression, i.e., production of hormones, described by Posner *et al.*, has been a productive approach to understand controls for predetermined expressions. Their report summarizes existing knowledge about clonally derived and permanently established cell cultures which maintain response to and/or production of special products or capabilities. These cell lines have been derived by serial explantation *in vitro* and transplantation *in vivo* accompanied by selection for capability to retain stabilized and differentiated syntheses.

The counterparts of hormone *producing* cell lines are those that are hormone *dependent*. Such colonies were obtained by culturing cells in the presence of bromodeoxyuridine and in the absence of stimulating hormones. A short exposure to fluorescent light caused self-destruction of those cells that, in the absence of hormone, had incorporated bromodeoxyuridine into their DNA. This left, as survivors, cells that failed to divide in the absence of the hormone, including those that may have been stimulated by or may have required the missing hormone. Thus, subsequent growth in culture in the presence of the hormone established cell lines which, in order to sustain their previously established capabilities, were dependent upon the hormone(s) missing during bromodeoxyuridine treatment.

The first procedure, serial explant and transplant, may have selected for, or locked in, effective gene copies for assuring cell multiplication and for effective production of the selected, specialized product for export. The second procedure, hormone withdrawal and self destruct, may have selected for cells which had effective gene copies for clonal and rapid multiplication concomitant with reduced or missing gene copies for the selected dependency. This exaggerated responsiveness to hormones apparently has enhanced the probability of selection for those cellular expressions dependent upon hormonal stimulation. Out of these procedures are derived cell strains in which cellular expression has been artificially amplified and in which there presumably is dominance of the effects of gene copies for continued expression of specific cell properties.

Urist describes extensive experiments characterizing the conditions under which mesenchyme cells differentiate within an explanted, three-dimensional morphogenetic matrix, i.e., substrate. First, amoeboid mesenchyme cells wander into the acellular matrix, attach and multiply for several days. Second, cells fuse into clusters and resorb the demineralized matrix. Three, differentiation occurs into blast cells, i.e., osteoblasts, hemocytoblasts, and chrondroblasts. During this sequence of changes the resorption cavity acquires extracellular calcifiable matrix with nucleation sites and protein cross-linkages. It acquires chondromucoproteins. It acquires blood forming marrow. Deposition of new bone, first seen at about 15 days after implantation of decalcified matrix, is at a distance from cells and between collagen fibers. Subsequent to calcification at the nucleation sites, the collagen fibers themselves become surrounded by bone deposition. Collagen adjacent to cells is excluded from the mineralization process.

The experiments of Urist parallel those of Bernfield and Wessells and of Dische by emphasizing that important antecedents to expression occur in an organized way when appropriate compounds in appropriate amounts are exported into extracellular spaces. The biological system reported by Urist appears to run the gamut of (a) cellular changes within an implanted substrate matrix, (b) cellular multiplication, (c) initiation of new cellular expressions, i.e., exported matrix products, (d) irrevokable commitment of mesenchymal cells, and (e) sustained terminal differentiations. Further study on the relationship between environmental influences and transcriptional and posttranscriptional events which maintain and control expresssions in this system will be most interesting.

SUMMARY

The Society for Developmental Biology, by means of a multidisciplinary symposium, once again has attempted to survey some of the perplexities so clearly expressed by E. B. Wilson. The introductory statements for this chapter, taken from Wilson (1902) out of context and out of sequence, are all the more remarkable because they were written before the biological world fully appreciated Mendel's contribution. These quotations were published the year that Sutton, working in Wilson's own laboratory, "pointed out clearly and completely that the known behavior of the chromosomes at the time of maturation of the germ cells (extensively reviewed by Wilson, 1902, Chapter V.) furnishes us with a mechanism that accounts for the kinds of separation of hereditary units postulated in Mendel's theory" (T. H. Morgan, 1935). Today's diversity of input has led to some uncertainty about what constitutes a developmental biologist. It remains to be seen which investigators in which facets of this multidisciplinary field will emerge as major contributors, hence leaders. Perhaps an operational criterion for a developmental biologist is that, in his own laboratory and classroom, he is one who relates his special contributions to changing syntheses during cell and tissue interactions in "higher" organisms.

It is with the aim of further stimulating multidisciplinary interaction that this chapter has attempted to weave the contents of the symposium into a working model (Fig. 1) for testing correlations between syntheses and successions of changes in cell and tissue interactions. Apparently we herald an era in which the identity with developmental biology contributes relevance, coherence, and mission for a multidisciplinary field.

REFERENCES

CALLAN, H. G. (1967). The organization of genetic units in chromosomes. *J. Cell Sci.* **2**, 1–7.

BRITTEN, R. J., and DAVIDSON, E. H. (1969). Gene regulation for higher cells. *Science* **165**, 349–357.

DENIS, H. (1968). Role of messenger ribonucleic acid in embryonic development. *In:* "Advances in Morphogenesis", Academic Press, New York. **7**, 115–150.

MORGAN, T. H. (1935). The relation of genetics to physiology and medicine. Nobel Prize Lecture June 4, 1934. *Sci. Mon.* **41**, 5–18.

WADDINGTON, C. H. (1967). "Principles of Development and Differentiation." Macmillan, New York.

WILSON, E. B. (1902). "The Cell in Development and Inheritance," 2nd ed. Macmillan, New York.

The Role of Gene Redundancy and Number of Cell Divisions in Embryonic Determination[1]

R. A. FLICKINGER

Department of Biology, State University of New York, Buffalo, New York 14214

INTRODUCTION

One purpose of this paper is to examine the number of copies and kinds of DNA-like RNA (D-RNA), i.e., the RNA with a base composition similar to DNA, that are synthesized in the nuclei and conserved in the cytoplasm during early development of the frog embryo, *Rana pipiens*. Correlations will be drawn between a restriction in kinds of D-RNA transcribed from redundant DNA in the nuclei during development and the corresponding restriction in embryonic competence, i.e., the potential pathways of differentiation open to the cells. During this period of restriction more copies and kinds of D-RNA accumulate and are conserved in the cytoplasm, and this will be related to embryonic determination, i.e., the setting of the fate of embryonic cells. Evidence will be presented which indicates that the restriction of transcription from the redundant DNA occurs because this DNA becomes late-replicating. The effect of the rate and number of cell divisions on the number of copies and kinds of D-RNA will be examined. Finally, experiments which relate gene redundancy and evolutionary age of the genes to the temporal sequence of D-RNA accumulation and cell determination will be presented.

RELATION BETWEEN RATE AND NUMBER OF CELL DIVISIONS AND RNA SYNTHESIS

The normal developmental sequence of competence and determination depends upon DNA synthesis and cell division (Flickinger, 1966). It is likely that the synthesis of D-RNA also depends upon DNA synthesis and cell division. Furthermore, the synthesis and conservation of D-RNA may account for sequential changes of competence and determination.

[1] This research was supported by grants from the National Science Foundation (GB5500) and the National Institutes of Health (GM 16236-01).

Amount of D-RNA Synthesized in Relation to Rate and Number of Cell Divisions

Number of cell divisions. From gastrulation to the early swimming larval stage of *Xenopus laevis*, the content of D-RNA doubles each time the DNA content of the embryo doubles (Brown and Littna, 1966). There is no increase in dry weight from fertilization to the feeding larval stage in developing embryos of *Rana pipiens*, and as increasing numbers of nuclei fill the embryo, there is an accumulation of D-RNA.

Rate of cell divisions. More slowly dividing endoderm cells of *Rana pipiens* (Flickinger *et al.*, 1967a, 1970a) and *Xenopus laevis* (Woodland and Gurdon, 1968) accumulate more total RNA *per cell*, including D-RNA, than do the ectoderm-mesoderm cells, which divide more rapidly. When rate of cell division of embryo explants is decreased by culture in LiCl, there is an increase in the amount of RNA accumulated *per cell*, inclucing D-RNA, (Flickinger *et al.*, 1967a). Increasing the rate of cell division by culture of embryo explants in NaHCO$_3$ causes a decrease in the amount of RNA synthesized *per cell*, including D-RNA, but more total RNA is accumulated because the explants underwent more cell divisions (Flickinger *et al.*, 1970a).

Relation of Cell Division to Differentiation

Ectoderm of amphibian gastrulae, which normally forms epidermis *in vitro*, will differentiate into endoderm and mesoderm in the presence of LiCl (Barth and Barth, 1962; Gehardt and Nieuwkoop, 1964; Masui, 1966; Ôgi, 1961). If NaHCO$_3$ is in the medium, neural differentiation occurs (Barth, 1966). The normal counterpart of the experimental neural induction by NaHCO$_3$ is illustrated by the increased cell division in the neural plate-dorsal mesoderm region of early neurulae, compared to the lateral epidermis-mesoderm regions (Flickinger *et al.*, 1970a)

Another link between cell division and determination is that endoderm cells divide more slowly than do the ectoderm-mesoderm cells and the fate of the endoderm cells is set after fewer cell divisions than the ectoderm-mesoderm cells (Flickinger *et al.*, 1967a). Hence, it is possible that the reduced rate and number of cell divisions due to LiCl leads to endoderm or mesoderm formation. Salamander gastrula ectoderm cells which undergo fewer divisions than usual will differentiate into endodermal derivatives in long-term cultures (Takata and Yamada, 1960).

Separation of Rate and Number of Divisions

The rate of cell division also was altered by raising embryos at a lower (10°–11°C) or higher (23°–24°C) temperature than the normal culture temperature at 20°C. At 23°–24°C it took 2–3 days for blastulae to develop to a tailbud stage and hatch from their membranes, and at 10°–11°C it took 8–9 days. Both groups of embryos were normal and possessed similar amounts of DNA per embryo. However, the embryos raised at the lower temperature had higher RNA:DNA ratios (3.09) than those raised at the higher temperature (2.48) (Flickinger et al., 1970b). Krugelis et al. (1952) have also found higher RNA:DNA ratios in salamander embryos raised at lower temperatures.

A hybridization procedure which allowed repeated withdrawals of hybridized RNA from the same RNA sample was used to find the percent of D-RNA present in the embryos raised at the two temperatures. The species of D-RNA present in fewer copies relative to their homologous DNA sequences were exhausted more rapidly than the species of D-RNA present in a greater number of copies; therefore this "exhaustion" method compares the number of copies of D-RNA synthesized from the redundant sequences of DNA.

Total RNA was extracted from tailbud embryos raised at 10°–11°C and 23°–24°C and labeled *in vitro* with dimethyl sulfate-^{3}H (Smith et al., 1967). Nitrocellulose filters, containing 15 µg of denatured DNA from red blood cells of frogs, were repeatedly incubated in solutions containing 20 µg of these two labeled RNA preparations. After each incubation the filters were washed, treated with RNase, and dried; the hybridized RNA was hydrolyzed and quantitated by isotope counting (Fig. 1). The results of three such experiments, together with the ratios for total RNA and DNA content, show the D-RNA:total DNA, or D-RNA per cell, of embryos raised at the lower temperature to be 33% ($\pm 4\%$) greater than the D-RNA:DNA ratio of embryos raised at the higher temperature (Flickinger et al., 1970b). It appears that a greater number of copies of D-RNA *per cell* were synthesized in the embryos raised at a lower temperature and presumably as a result of their slower rate of cell division.

The time of determination of the lens (Ten Cate, 1953) and ciliary polarity (Twitty, 1928) occur at an earlier morphological stage of devel-

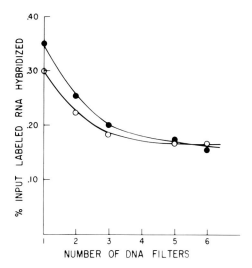

Fig. 1. The depletion of hybridizable labeled RNA from 20 μg RNA samples by the addition of successive nitrocellulose filters containing 15 μg of denatured frog red blood cell DNA is shown for RNA isolated from embryos raised at 10–11°C (●) and 23°C (○).

opment in amphibian embryos raised at a temperature lower than normal. Since the rate of division is decreased at the lower temperatures, it is possible that this accounted for precocious determination. Similarly, redirection of prospective pathways of differentiation by temperature changes has been demonstrated for the imaginal discs of *Drosophila* larvae (Villee, 1944).

Apparent contradictory results between rate division and D-RNA synthesis per cell.

Although more slowly dividing endoderm cells accumulate more labeled D-RNA *per cell* than do the more rapidly dividing ectoderm and mesoderm cells, they become determined and differentiate later in time (Balinsky, 1948). A possible explanation is that less of the D-RNA accumulated in the cytoplasm because there are fewer ribosomes in the endoderm cell cytoplasm (Brachet, 1965) to "conserve" the D-RNA as polysomal D-RNA which would be resistant to ribonuclease and not be hydrolyzed. A short half-life for D-RNA in endodermal cells, compared to the dorsal ectoderm-mesoderm cells, has been demonstrated by actinomycin D experiments with neurula and tailbud embryos (Flickinger, 1970).

A more direct demonstration of the effect of the number of monosomes on "conservation" of newly synthesized D-RNA was attempted.

Ribosome preparations from oocytes injected into vegetal blastomeres at the 8-cell stage caused an increase in accumulation of labeled RNA (Rollins and Flickinger, unpublished data). Thus, it is possible that D-RNA in more slowly dividing cells may induce precocious determination in the presence of a greater number of ribosomes which conserve more of the D-RNA.

Kinds of D-RNA Synthesized in Relation to Rate and Number of Cell Divisions

Both the rate and number of cell divisions is influenced when embryo explants are cultured in $NaHCO_3$ and LiCl for the same period of time. It was possible to compare the effect of an increase in rate of division alone by culturing neurula explants in $NaHCO_3$ for a shorter period of time so that they underwent a similar number of cell divisions as explants cultured in LiCl for a longer period. Since LiCl and $NaHCO_3$ can alter pathways of differentiation, it is critical to ascertain whether these compounds induce qualitative, as well as quantitative, differences in RNA synthesis. DNA-RNA hybridization experiments were employed for this purpose. RNA synthesized *in vitro* from isolated chromatin was used so that the amounts and kinds of D-RNA synthesized would not depend upon the number of cell divisions during incubation in labeled uridine. Saturation curves, i.e., the successive additions of labeled RNA until no more labeled RNA hybridizes with denatured DNA, were obtained for the hybridizable RNA-^{3}H produced *in vitro* (using added microbial RNA polymerase) with chromatin isolated from neurula explants treated 24 hours with LiCl or $NaHCO_3$, or treated for 13 hours with $NaHCO_3$ (Flickinger et al., 1970c). The RNA produced from chromatin isolated from the 24-hour-$NaHCO_3$ embryos had a lower level of saturation than the RNA transcribed from chromatin of embryos cultured in LiCl for 24 hours or $NaHCO_3$ for 13 hours (Fig. 2). This indicates that fewer kinds of D-RNA were synthesized by the chromatin of the 24 hour-$NaHCO_3$ embryos and that less of the genome was active. The RNA that hybridizes reflects that transcribed from redundant DNA sequences only (Britten and Kohne, 1968). The chromatin from 13-hour-$NaHCO_3$ and 24-hour-LiCl embryos synthesize more kinds of D-RNA than the chromatin from 24-hour-$NaHCO_3$ explants and this is believed to be because the 13-hour $NaHCO_3$ and the 24-hour-LiCl cultures are mitotically younger.

These data show that, as mitotic age increases, transcription is re-

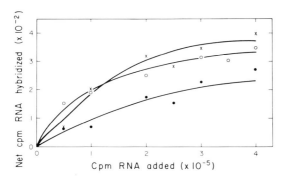

FIG. 2. Saturation curves for labeled RNA synthesized *in vitro* from 1 mg of chromatin DNA isolated from embryos treated with 0.25% LiCl in Niu-Twitty saline for 24 hours (×) and with 0.25% NaHCO$_3$ in Niu-Twitty saline for 13 (○) and 24 hours (●). *Micrococcus lysodeikticus* RNA polymerase and other components of the system were present in 10 times their usual concentration.

stricted qualitatively in each cell during this period of frog embryo development. However, a slower rate of division allows the synthesis of more copies of the D-RNA molecules that are capable of being synthesized at that particular mitotic age. Although the slowing of rate of division is important, LiCl and NaHCO$_3$ may direct differentiation by altering the number of cell divisions in isolated gastrula ectoderm.

Accumulation of More Kinds of D-RNA during Development

As *Xenopus* embryos develop from gastrulae to swimming larvae hybridization experiments reveal that more kinds of D-RNA accumulate in the embryos (Denis, 1966). In our laboratory total RNA has been prepared from different stages of developing embryos of *Rana pipiens* and labeled *in vitro* using dimethyl sulfate-^{3}H (Smith *et al.*, 1967). These labeled RNA preparations from whole embryos were used for DNA-RNA hybridization experiments (Greene and Flickinger, 1970). These labeled RNA preparations were hybridized to saturation with denatured red blood cell DNA on nitrocellulose filters. Saturation experiments show an increase in the percent of the genome hybridized from a level of 1.5% for late gastrulae (Fig. 3a) to 8% for swimming larvae (Fig. 3d), indicating that more kinds of D-RNA have accumulated during this period of development.

Competition experiments, i.e., the simultaneous addition of unlabeled RNA together with the labeled RNA in the hybridization

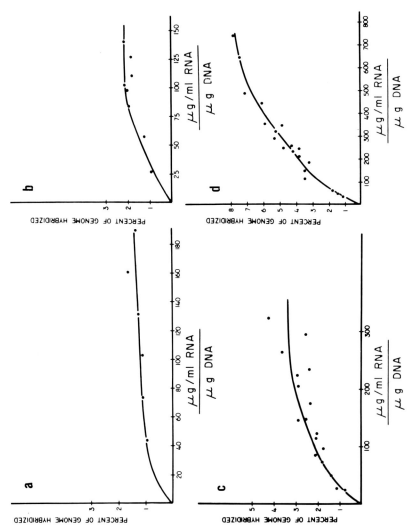

FIG. 3. Saturation curves: RNA was isolated from developing *Rana pipiens* embryos, labeled *in vitro* with dimethyl sulfate-^{3}H, and incubated for 48 hours in 2 × SSC (I × SSC is 0.15 M NaCl, 0.15 M sodium citrate), 50% formamide at 37°C with denatured frog red blood cell DNA immobilized on nitrocellulose filters. (a) Composite curve from two separate experiments with gastrula (stage 12) RNA; the filters contained 3 μg of DNA. (b) Composite curve of three separate experiments with RNA from neurulae (stage 14); the filters contained 5.1 and 3 μg of DNA. (c) Composite curve of four separate experiments with tailbud (stage 18) RNA; filters contained 5.6, 3, and 1.5 μg of DNA. (d) Composite curve of four separate experiments with RNA from larvae (stage 25); filters contained 5.1, 3, and 1.5 μg of DNA.

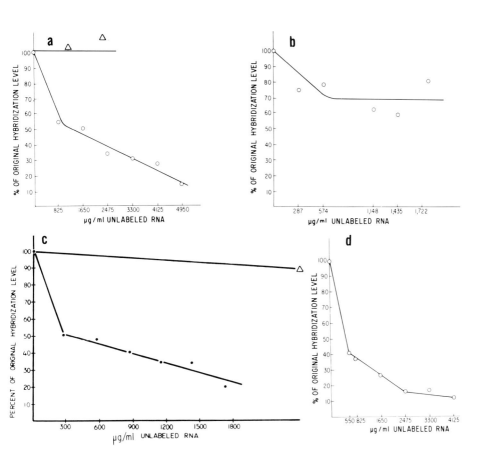

FIG. 4. Competition experiments: (a) competition of unlabeled larval (stage 25) RNA against homologous labeled larval RNA (○). Increasing amounts of unlabeled RNA were competed against 720 µg/ml of labeled RNA. DNA filters contained 1.5 µg of denatured DNA. Yeast RNA (△) as competitor also was used. (b) Competition of unlabeled neurula (stage 14) RNA against labeled larval (stage 25) RNA. Increasing amounts of labeled RNA was competed against 720 µg/ml of labeled RNA. DNA filters contained 1.5 µg of denatured DNA. (c) Competition of unlabeled neurula (stage 14) RNA against homologous labeled neurula RNA. Increasing amounts of unlabeled RNA (●) were competed against 290 µg/ml of labeled RNA. DNA filters contained 3µg of denatured DNA. Yeast RNA competitor (△). (d) Competition of unlabeled larval (stage 25) RNA against labeled neurula (stage 14) RNA. Increasing amounts of unlabeled RNA were competed against 290 µg/ml of labeled RNA. DNA filters contained 3 µg of denatured DNA.

incubation, also reveal that new species of D-RNA progressively accumulate during development (Fig. 4). "Exhaustion" experiments were also performed in which nitrocellulose filters containing denatured DNA were successively incubated with the same sample of labeled RNA. If the ratios of certain species of D-RNA to their DNA cistrons vary, the exhaustion experiments will reveal this. It is apparent that there is a class of D-RNA which is rapidly depleted from the labeled larval RNA preparations and that this is not true for the labeled neurula RNA in which only a small amount of RNA is rapidly depleted (Fig. 5). The RNA molecules which cannot be depleted in these experiments are ribosomal RNA, transfer RNA, and those D-RNAs which are present in great excess compared to their complimentary DNA sequences. These results suggest that there is a class of D-RNA present at the larva stage that is present in few copies relative to its homologous DNA sequence and this class of D-RNA is nearly absent in neurulae. However, another explanation is that a greater percentage of the D-RNA accumulating by the larval stage has been transcribed from more redundant sequences of the genome than at the neurula stage, since redundancy would reduce the ratio of certain species of D-RNA to their homologous DNA cistrons.

In comparing the rate of hybridization of neurula and larval RNA to denatured red cell DNA with time, Fig. 6 shows that the labeled larval RNA hybridizes slightly less rapidly. This indicates that the product of the RNA concentration times the DNA concentration is less at the larval stage than for neurulae, showing most of the D-RNA detected at the larval stage is not transcribed from more redundant sequences. This failure to find more transcription from redundant DNA by the larval stage suggests that there is a class of D-RNA present in few copies in intact larvae which is almost absent in intact neurulae.

Nuclear Synthesis of Different Species of D-RNA

In order to distinguish the number of kinds of D-RNA being synthesized in the nuclei at different stages of development, nuclei were isolated from neurulae, tailbuds, and larvae of *Rana pipiens*. RNA was isolated and labeled *in vitro*, and hybridization experiments were performed. At saturation levels labeled neurula nuclear RNA hybridized with 32% of the genome, labeled tailbud nuclear RNA with 10% of the genome, and labeled larval nuclear RNA with 4% of the genome (Fig. 7, Daniel and Flickinger, in press). It is apparent that

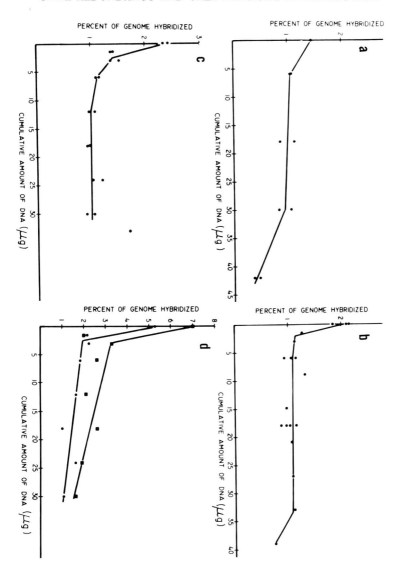

FIG. 5. Exhaustion experiments: labeled RNA was exhausted by incubation for 24 hours with consecutive DNA filters. (a) Gastrula RNA (stage 12). (b) Neurula RNA (stage 14). (c) Tailbud RNA (stage 18). (d) Larval RNA (stage 25). Two different concentrations of labeled larval RNA were used: 481 μg/ml (●) and 962 μg/ml (■).

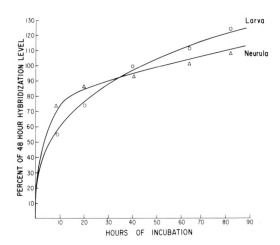

FIG. 6. Kinetic experiments: rate of annealing of neurula (stage 14) and larval (stage 25) RNA with time. RNA from neurulae (△, 890 µg/ml) and from larvae (○, 1040 µg/ml) were incubated with filters containing 5.1 µg of denatured DNA.

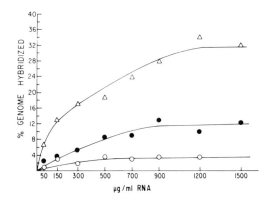

FIG. 7. Percent of genome hybridized by nuclear RNA of neurulae (△), tailbuds (●) and larvae (○) which had been labeled *in vitro* with dimethyl sulfate. The RNA samples were incubated with nitrocellulose filters containing 1.24 µg of denatured frog red blood cell DNA for 48 hours at 37°C in 2 × SSC and 50% formamide.

much of the genome is active at an early neurula stage and that there is a decrease in the number of kinds of D-RNA present in nuclei during development from an early neurula to a swimming larval stage. The apparent contradiction between the increase in number of kinds of D-RNA in whole cells (Fig. 3) and the decrease in number of kinds of D-RNA in nuclei (Fig. 7) during the same period of development is attributed to transport and conservation of more kinds of D-RNA in the cytoplasm during development. The hybridization method detects only D-RNA transcribed from redundant DNA sequences, and the decreasing levels of saturation with nuclear RNA of later stages means that there is a reduction in transcription from redundant DNA from the neurula to the larval stage.

A series of competition experiments also were performed with the labeled nuclear RNA preparations (Daniel and Flickinger, in press). As expected the addition of unlabeled larval nuclear RNA offers little competition to the hybridization of labeled neurula nuclear RNA, while unlabeled neurula RNA competes as well as unlabeled larval nuclear RNA when labeled larval RNA is hybridized to denatured DNA (Fig. 8a). This means that all the kinds of D-RNA transcribed from redundant DNA which are present in larval nuclei are also present in neurula nuclei. Unlabeled neurula nuclear RNA also competes as well as unlabeled RNA from whole larvae when labeled whole larval RNA is hybridized (Fig. 8b). This shows that neurula nuclei contain all the kinds of D-RNA present in intact swimming larvae. Furthermore, unlabeled neurula nuclear RNA competes almost as well as unlabeled adult frog liver RNA when labeled liver RNA is hybridized to DNA (Fig. 8c). This implies that neurula nuclei contain numerous representatives of the kinds of D-RNA present in adult frog liver.

Our hybridizations were performed using the method of Gillespie and Spiegelman (1965), and D-RNA synthesized from unique sequences of DNA is not detected in our experiments (Britten and Kohne, 1968). It is possible that the D-RNAs examined are not messenger-RNA, or if they are, they are messenger RNAs coding for enzymes and structural proteins present in all frog cells. However, the fact that the hybridization procedure used does detect pronounced changes in the composition of the D-RNA species in nuclei and whole cells during the critical period of early embryonic development, when determination and differentiation occur, suggests that such changes may play a role in deciding what pathway of determination the cells will follow.

Near-saturating levels of labeled nuclear RNA preparations of neurulae, tailbuds, and larvae were exposed to successive incubations with filters containing denatured DNA. The prompt and greater degree of exhaustion of labeled neurula nuclear RNA by successive DNA filters means that the neurula nuclei contained more representatives of this class of rapidly exhausted D-RNA than did tailbud nuclei, which

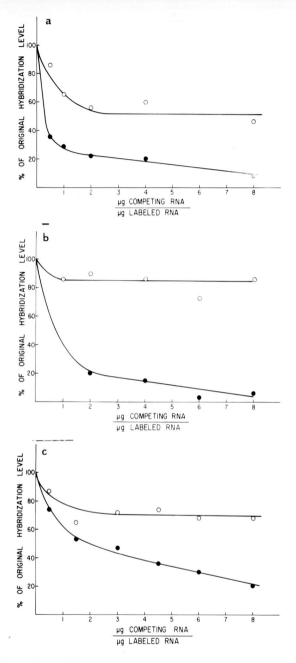

FIG. 8. Competition of (a) unlabeled stage 25 larval nuclear RNA with labeled stage 14 neurula nuclear RNA (O); unlabeled stage 14 neurula nuclear RNA with labeled stage 25 nuclear RNA (●). (b) Unlabeled RNA of intact stage 25 larvae with labeled stage 14 neurula nuclear RNA (O); unlabeled stage 14 neurula nuclear RNA with labeled RNA of intact stage 25 larvae (●). (c) Unlabeled RNA of adult frog livers with labeled stage 14 neurula nuclear RNA (O); unlabeled RNA of stage 14 neurula nuclei with labeled RNA of adult frog livers (●).

in turn contained more members of this class of D-RNA than did larval nuclei (Fig. 9). The higher plateau level for the neurula nuclear RNA implies there are more species of the class of D-RNA present in many copies relative to their homologous DNA sequences in neurula nuclei than in nuclei of later stages. Quantitatively there is reduced transcription of both of these classes of D-RNA from the neurula to the larval stage (Fig. 9). The course of hybridization of labeled neurula and larval nuclear RNA with time is similar, implying that there is not a proportionately greater transcription from more redundant DNA at the earlier stage. Exhaustion experiments with labeled RNA of intact embryos showed there were more representatives of the class of rapidly exhausted D-RNA at later stages of development (Fig. 5). This is attributed to accumulation and conservation of D-RNA in the cytoplasm with development.

RELATION OF CELL DIVISION TO A PARTICULAR PATHWAY OF DETERMINATION

If the number of divisions allows accumulation of particular D-RNA molecules, this would be expected to occur in all cells that were competent to undergo the determination characterized by these D-RNA molecules. For example, it is known that any region of the flank ectoderm of *Rana pipiens* tailbud embryos can be induced to form a lens

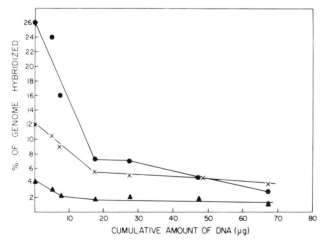

FIG. 9. The exhaustion of 600 μg/ml labeled RNA of stage 14 neurula nuclei (●), stage 18 tailbud nuclei (×) and stage 25 larval nuclei (▲) was measured by successive incubations with filters containing denatured DNA.

(Liedke, 1942). Some of these D-RNA molecules may be translated as shown by the presence of lens antigens in the flank regions (Clayton, 1968; Flickinger and Stone, 1960). Furthermore, when gastrulae are exposed to a high concentration of ethionine (0.6%) for a day, then allowed to develop further in the absence of this amino acid analog, abnormally placed multiple lenses form (Flickinger, 1961). Earlier determination of the lens occurs in embryos raised at a colder temperature (Ten Cate, 1953). If determination corresponds to a cytoplasmic accumulation and conservations of sufficient D-RNA responsible for the determination of the lens, then since the length of the cell cycle is increased at the lower temperature, the determination of the lens after fewer cell divisions implies that the slower rate of division might allow an earlier accumulation of this D-RNA. The exhaustion experiments performed with labeled RNA obtained from embryos raised at 10°–11°C and 23°–24°C shows that more copies of D-RNA are transcribed or conserved in embryos raised at 10°–11°C than those raised at 23°–24°C (Fig. 1), even though the number of cell divisions is similar. If the increase in number of copies of D-RNA also raises the number of D-RNA molecules responsible for the determination of lens above a critical threshold, this could explain the earlier determination of the lens. It is also possible that the production of more ribosomal RNA at the lower temperature (Flickinger

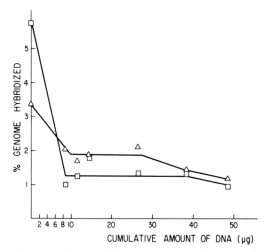

Fig. 10. The exhaustion of labeled RNA of stage 19 tailbuds raised at 10°–11°C (△) and stage 19+ tailbuds raised at 23°–24°C (□) was measured by successive incubations with filters containing denatured DNA.

et al., 1970b) could account for the conservation of more of the D-RNA responsible for lens determination.

If the number and rate of cell divisions can control the process of embryonic determination, this may explain why it is that the agents which modify pathways of cell determination are in themselves not highly specific. Aside from obvious nonspecific means such as animalizing-vegetalizing agents or ethionine, it should be emphasized that embryonic inductors lack specificity (Holtfreter and Hamburger, 1955). Also the inhibition of a particular determination at its normal site with metabolic inhibitors will allow another region which normally will not follow that pathway of differentiation to do so (Flickinger and Coward, 1962). This implies that these other regions are capable of accumulating the specific D-RNAs for these determinations. Alteration of differentiation by metabolic agents usually affects the number and rate of cell divisions which in turn may influence embryonic determination.

POSSIBLE CONTROL MECHANISMS FOR TRANSCRIPTION AND DETERMINATION

Progressively fewer redundant genes are transcribing D-RNA during development from the neurula to the larval stage as shown by hybridization experiments with labeled nuclear RNA (Fig. 7). The increase in kinds of D-RNA accumulating during development (Denis, 1966) appears to be accounted for by increased accumulation and conservation of D-RNA in the cytoplasm during this period. The most apparent questions arising from these experiments are: What turns genes off during early development? What accounts for selective accumulation in the cytoplasm?

Rate of Cell Divisions and Qualitative and Quantitative Changes in D-RNA Synthesis

A slower rate of cell division is associated with increased D-RNA synthesis *per cell*, and a more rapid rate with decreased D-RNA synthesis *per cell* (Flickinger *et al.*, 1970a). This possibly could operate to control D-RNA synthesis qualitatively since fewer kinds of D-RNA are synthesized from the neurula to the larval stage (Fig. 7). A slower rate of cell division would favor the transcription of more copies of those D-RNAs being synthesized at that time, while a rapid rate of cell division would tend to allow transcription of fewer copies of those D-RNA molecules (Fig. 1).

However, it appears to be impossible to dissociate DNA synthesis and cell division from the normal time course of changing competence, i.e., the changes in the potential of various kinds of differentiation. An inhibitor of DNA synthesis, cytosine arabinoside, caused a reversible delay of normal development of frog gastrulae (Flickinger, 1966) with no inhibition of RNA or protein synthesis. This suggests that competence changes with cell division. The mitotic age of the cells might control what kinds of D-RNA can be synthesized, while the rate of division could control the number of copies transcribed.

The exhaustion experiments performed with RNA obtained from embryos raised from the blastula to the tailbud stage at 10°–11°C and 23°-24°C, together with the total RNA:total DNA ratios, indicate that more copies of D-RNA are synthesized and conserved in the embryos raised at a colder temperature (Fig. 1). Since these embryos raised at these two temperatures underwent a similar number of cell divisions, as shown by determinations of total DNA, it appears that quantitative differences in D-RNA synthesis result from variations in the rate of cell divisions.

In order to learn whether the production of D-RNAs which are present in fewer copies might raise the number of molecules from a level previously undetected by the hybridization experiment to a detectable level, saturation experiments were performed with RNA from embryos raised at 10°–11° and 23°–24°C. These RNA preparations were labeled *in vitro* with dimethyl sulfate-^{3}H (Smith *et al.*, 1967) and saturation experiments were performed. At saturation, both of the labeled RNA preparations of embryos that were raised to the same morphological stage (stage 19) and DNA level at 10°–11°C and 20°C hybridized with 4.5% of the DNA (Flickinger *et al.*, 1970b). This similar saturation pleateau indicated that the number of kinds of D-RNA present in both RNA preparations is similar in spite of the fact that the embryos raised at the lower temperature have more total RNA and D-RNA per cell (Fig. 1).

Labeled RNA of older stage 19+ tailbuds, raised from the blastula stage at 23°–24°C for 2–3 days, hybridized with 7% of the DNA at saturation. This is to be compared to 4.5% for the younger stage 19 tailbuds raised from the blastula stage at 10°–11°C for 8–9 days. These same labeled RNA preparations were used in exhaustion experiments at levels near saturation (1100 µg/ml). Consecutive DNA filters were added for 24 hours each and the labeled RNA which hybridized was counted. Although the younger stage 19 embryos raised at 10°–11°C

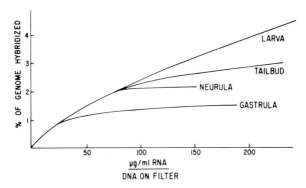

FIG. 11. The initial portions of the saturation curves of RNA isolated from gastrulae (stage 12), neurulae (stage 14), tailbuds (stage 18) and swimming larvae (stage 25). The RNA was labeled *in vitro* with dimethyl sulfate-^{3}H and then incubated for 48 hours in $2 \times$ SSC-50% formamide at 37°C with denatured frog blood cell DNA immobilized on nitrocellulose filters.

had fewer kinds of D-RNA than the older tailbud embryos (stage 19+) raised at 23°–24°C, the exhaustion experiment shows that the younger embryos raised at 10°–11°C over a longer period of time have more copies of D-RNA (Fig. 10). These experiments demonstrate that a slower rate of division allows more copies of D-RNA to be conserved, but that more kinds of D-RNA accumulate in embryos which underwent a greater number of cell divisions.

Accumulation of the Class of D-RNA Present in Few Copies

Britten and Kohne (1968) speculated that the degree of redundancy of certain cistrons could be a factor which controlled the synthesis of RNA molecules. The DNA-RNA saturation experiments performed with whole embryo RNA support the contention that most of the D-RNA detected at the gastrula and neurula stages is present in many copies since saturation is attained at lower input ratios of labeled RNA to DNA than for later stages (Fig. 11). By the tailbud and larval stages the slopes of the saturation curves are decreased and more total labeled RNA is needed to reach saturation (Fig. 11). It may be that cell division allows a change in embryonic competence by the accumulation of more copies of the less numerous class of D-RNA with successive cell divisions.

In the cells which are determined after fewer cell divisions, cessation of cell division after determination might prevent a sufficient accumulation of the genes which synthesize the less numerous D-RNA molecules which would account for later types of competence. If some of the D-RNA species present in many copies code for the synthesis of proteins common to all cells, then a selective reduction in transcription of these molecules would be unnecessary in order for the D-RNA present in fewer copies to control determination.

At the blastula and gastrula stages of developing fish embryos, transcription is primarily from redundant sequences of DNA (Rachkus et al., 1969a, b). As development proceeds there is a decrease in transcription of the repeating DNA sequences, and these workers speculate that there is an increase in transcription from less redundant regions of the genome. While our exhaustion and kinetic experimenting partly agree with this possibility, we are currently engaged in hybridizing frog DNA of varying degrees of redundancy to labeled RNA of early and late-stage frog embryos in order to test this idea in a conclusive manner. Evidence from hybridization experiments with labeled nuclear RNA indicates reduced transcription from redundant DNA during early development of the frog embryo (Fig. 7).

Late Replication of DNA as a Means of Restricting Transcription

One way in which transcription from certain DNA cistrons could be restricted during development is that these DNA sequences would become late replicating (Hsu et al., 1964) and thereby limit the time per mitotic cycle for transcription. A determination of the amount of DNA that replicates late in the S period has been made for *Rana pipiens* embryos by autoradiography (Stambrook and Flickinger, 1970) and by using partially synchronized cells for biochemical estimates of DNA synthesis (Remington and Flickinger, in press).

Autoradiographic experiments. Patterns of late-replicating DNA were obtained by incubating parts of gastrulae, neurulae, and tailbuds in thymidine-^{3}H for a period equivalent to the duration of the G_2 period plus the last quarter of the S period; both periods had been previously determined (Flickinger et al., 1967b, 1970a). It appears that at the early gastrula stage, when most genes are still relatively inactive, many of the chromosomal DNA sequences are replicating during the last 15–30 minutes of the 1.5-hour S period (Fig. 12a). The neural plate-dorsal mesoderm and endoderm of neurulae show few DNA sequences replicating late in the 5–6-hour S period (Fig. 12b). The differentiated tailbud dorsal ectoderm-mesoderm shows numerous sites of late-replicating DNA (Fig. 12c). The autoradiographic data show that tailbuds have more late-replicating DNA than do neurulae. This correlates well with our hybridization data which reveals that tailbud embryos transcribe less of the genome than do neurulae (Fig. 7).

These changes in time of DNA replication occur at a certain stage, rather than being associated with a discrete population of cells within any given region. No differences in patterns of late-replicating DNA were observed for the same karyotyped chromosome in cells that undergo different pathways of specialization, e.g., neural plate and neurula endoderm. The onset of early replication of DNA between gastrulation and neurulation is correlated with an increase in the kinds of D-RNA synthesized (Denis, 1966; Kohl et al., 1969). It appears that stages synthesizing more kinds of D-RNA (neurulae) may

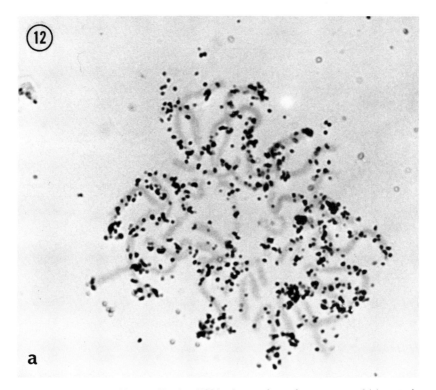

FIG. 12. Regions of late-replicating DNA of metaphase chromosomes of (a) gastrula ectoderm, (b) neural plates, and (c) the dorsal axial region of tailbud embryos.

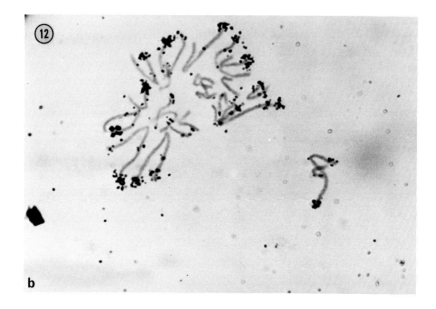
b

have more early-replicating sequences of DNA than stages that make fewer kinds of D-RNA (gastrulae and tailbuds).

Biochemical experiments with partially synchronized cells. Differences in the amount of late-replicating DNA in the dorsal axial and belly endoderm regions of neurulae and tailbud embryos were studied biochemically also. The cells of isolated neural plate-dorsal mesoderm and belly endoderm regions of neurulae and the dorsal axial regions (nerve tube, somites, notochord) and belly endoderm regions of tailbud embryos were partially synchronized with 5-fluorodeoxyuridine treatment and thymidine rescue (Remington and Flickinger, in press). The explants were cultured in Niu-Twitty saline (1953) containing 10^{-5} M 5-fluorodeoxyuridine for 14 hours in order to accumulate a population of cells blocked at the beginning of the S period. This block was removed by the addition of excess unlabeled thymidine. The degree of synchrony obtained according to the increase in the number of nuclei labeled autoradiographically was approximately 27% for axial parts and 46% for belly parts. The degree of synchrony obtained according to the biochemical criterion of the increase in incorporation of deoxyguanosine-^{3}H into the DNA of partially synchronized cells was approximately 34% for axial parts and 44% for belly parts. The S periods of the 5-fluorodeoxyuridine released and control cells at both the neurula and tailbud stages were approximately 8 hours. Thymidine-^{3}H, deoxyguanosine-^{3}H, and 5-uridine-^{3}H were used to measure DNA and RNA synthesis. The labeled compounds were added for 1- or 2-hour periods at intervals of 2 hours over a period of 12 hours. Two discrete peaks of DNA synthesis were observed (Figs. 13a and 13b), and they are believed to represent DNA that is replicating early and late in the S period, respectively.

The amount of labeled DNA reveals that at the tailbud stage the percentage of DNA replicating during the last third of the S period is greater than at the neurula stage in both the dorsal axial and belly endoderm regions (Figs. 13a and 13b). It is also apparent that endoderm possesses more late replicating DNA than dorsal ectoderm-mesoderm of the two stages. Experiments with labeled tailbud nuclear RNA of these regions shows that the endoderm with the most

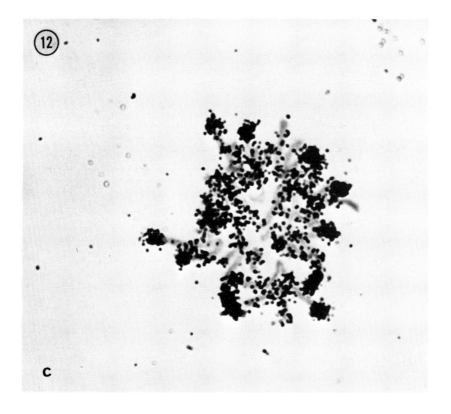

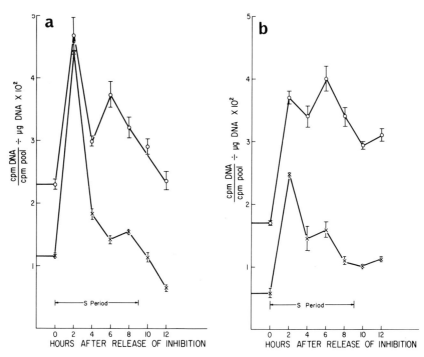

FIG. 13. (a) Groups of 50 neural plates (×) and belly parts (○) were incubated for 14 hours with 10^{-5} M 5-fluorodeoxyuridine to inhibit DNA synthesis. The inhibition was released with 10^{-4} M thymidine, and DNA synthesis was measured by 2-hour pulses of deoxyguanosine-^{3}H (2 μCi/ml) at 2-hour intervals. (b) The same procedure as in 13a was used to measure DNA synthesis in dorsal axial (×) and belly regions (○) of tailbud embryos.

late-replicating DNA makes fewer kinds of D-RNA (unpublished data).

The RNA precursor 5-uridine-^{3}H was given in 1-hour pulses during the S period of the partially synchronized neurula and tailbud explants. In neurulae the RNA is synthesized throughout the S period, but in parts of tailbuds the RNA synthesis during the last half of the S period is reduced (Fig. 14). These results imply that the initiation of late-replication of DNA is a means by which transcription is reduced.

EVOLUTIONARY IMPLICATIONS OF THE ROLE OF GENE REDUNDANCY IN CELLULAR DETERMINATION

It has been postulated that the number of cell divisions is a control mechanism for cellular determination. This control could operate by accumulating a sufficient number of genes which will ensure cytoplasmic conservation of the D-RNA molecules responsible for setting the fate of the cells. The effectiveness of gene dosage for accumulation of D-RNA is illustrated by the synthesis of twice as much RNA by cells of diploid fish embryos compared to cells of haploid embryos (Kafiani et al., 1968).

It is possible that transcription of more redundant DNA sequences is primarily responsible for determination of cells whose fate is set after fewer cell divisions. The reduction of number of divisions of gastrula ectoderm by LiCl, together with a slower rate of divisions, results in formation of endoderm or mesoderm. This might allow transcription of more redundant genes, while the less redundant genes would not accumulate in sufficient numbers. An increase in number of divisions by $NaHCO_3$, together with a more rapid rate of division, causes neural differentiation. With further divisions the less redundant genes would accumulate and their transcription would result in neural determination.

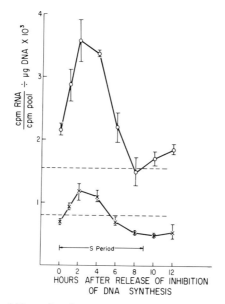

Fig. 14. Groups of 50 cut dorsal axial (×) and belly regions (○) of tailbud embryos were cultured 14 hours in 10^{-5} M 5-fluorodeoxyuridine to inhibit DNA synthesis. The inhibition was released with 10^{-4} M thymidine. RNA synthesis was measured by 1-hour pulses of 5-uridine-^{3}H (10 μCi/ml). The dotted lines indicate the levels of RNA synthesis in 5-fluorodeoxyuridine.

Additional mechanisms appear to be superimposed upon the primary mechanism for cell determination. If D-RNA molecules fail to be conserved in the cytoplasm or if D-RNA synthesis is restricted by late-replication of the more redundant DNA, then the potential for early determinations controlled by more redundant genes would be bypassed.

Whiteley et al. (1966, 1970) demonstrated two classes of D-RNA in developing sea urchins. One class was present in many copies in the unfertilized egg and embryo, and the synthesis of this class of D-RNA continued to the early pluteus stage. This D-RNA is thought to code for proteins used for cellular maintenance. Between hatching and gastrulation new kinds of D-RNA appear which are present in relatively few copies, and these are thought to control cellular differentiation. The D-RNA present in many copies is quite similar in various species of echinoids and asteroids, and the genes accounting for its synthesis are believed to be phylogenetically older. The D-RNA molecules appearing later in development are thought to be transcribed from genes that arise later in evolution. These gene differences may have occurred during evolution by repeated duplication of certain genes and may be reflected in the present-day genome by the differences in degree of redundancy. It is possible that the phylogenetically older genes may be more redundant than the younger ones.

The idea that older genes are more redundant finds support from the work of Britten and Kohne (1965), who compared the DNA of the mouse and the calf, as well as the mouse and the rat and found that there was a much greater percentage of redundant DNA held in common between these species than there were common sequences of unique DNA. Although the phylogenetically older redundant genes had undergone a good deal of divergence in their families, about 20% of the repeated sequences were common to the mouse and the calf, while few, if any, nonrepeated sequences were held in common. These workers do stress that some highly redundant DNA which has identical copying of its families has arisen later in evolution. Furthermore, they speculate that certain nonrepeated DNA sequences are the oldest genes phylogenetically and that repeating sequences arise by repeated duplications, yet they find that no non-repeating sequences are held in common between mouse, calf, and salmon DNA (Britten and Kohne, 1965). Since maintainance proteins such as cytochrome C, dehydrogenases, and phosphatases arose earlier in evolution than proteins characteristic of differentiated cells, it is likely that the genes responsible for cell maintenance were acquired earlier in evolution than those accounting for specialized differentiated functions. If older genes are more redundant, then it follows that redundant genes are more active in transcribing D-RNA involved in synthesizing proteins involved in cell maintenance. An example of this situation is seen when 5-bromodeoxyuridine causes the loss of specialized synthesis characteristic of differentiated cells, but does not interfere with the synthesis of metabolic products needed for cell survival (Abbott and Holtzer, 1968. If D-RNA responsible for differentiation is primarily transcribed from less redundant genes, while D-RNA accounting for cell maintenance is transcribed from more redundant DNA sequences, then a random minimal incorporation of 5-bromodeoxyuridine into DNA would result in a greater percentage of faulty 5-bromodeoxyuridine DNA in the less redundant sequences and result is differential reduction in special synthesis.

The detection of collagen synthesis at the early gastrula stage in *Xenopus* embryos (Green et al., 1968) and our demonstration that there is a low level of collagen syntheses in the endoderm cells of frog embryos (Klose and Flickinger, in press) predict that redundant genes are phylogenetically older. Since collagen has been demonstrated in the lower phyla, in fibers of sponges and coelenterates (Gross, 1963), there is added reason to predict that collagen synthesis depends upon D-RNA transcribed from redundant genes which were acquired early in phylogeny. It is obvious that the endocrine organs, bone, hair, and mammary glands, as well as proteins, such as insulin, thyroglobulin, ossein, keratin, and casein, which appear later in phylogeny, also appear later during the course of ontogeny after a greater number of cell divisions has occurred. It is possible that building up the number of genes of different degrees of redundancy in developing embryo by cell division may be the means by which "ontogeny recapitulates phylogeny."

While more redundant and evolutionarily older DNA sequences might account for the transcription of D-RNA responsible for cell determinations occurring early in ontogeny and phylogeny, in order for D-RNA transcribed from less redundant and evolutionarily younger DNA sequences to permit cell determinations which occur later in development and evolution, the early potentialities must be bypassed. This result would be obtained if the more redundant and phylogenetically older DNA sequences are more likely to become late-replicating, and hence transcribe less D-RNA than the less redundant DNA sequences. This possibility currently is under study in our laboratory.

SUMMARY

DNA-RNA hybridization experiments show that there are two general classes of DNA-like RNA (D-RNA); one class is present in many copies, and the other in few copies relative to their homologous DNA sequences. It is speculated that the DNA sequences transcribing D-RNA present in many copies are more redundant and evolutionarily older than the DNA sequences transcribing D-RNA present in few copies. All the kinds of D-RNA transcribed from redundant DNA that are present in swimming larvae are present in early neurula nuclei. There is a restriction of the kinds of D-RNA synthesized during this period of development, however, more kinds of D-RNA accumulate in the embryo due to the great increase in number of nuclei in the same amount of cytoplasm, resulting in a quantitative increase in the class of D-RNA present in few copies. A more efficient conservation of D-RNA in the cytoplasm contributes to the accumulation of more kinds of D-RNA in whole embryos as well. An increase in the quantity of late-replicating DNA is believed to account for the restriction of transcription during development.

The restriction of embryonic competence, i.e., the potential pathways of differentiation open to the cells, during development is ascribed to the restriction in kinds of D-RNA synthesized from redundant DNA during early development. Determination, i.e., the setting of the fate of a cell, is ascribed to the accumulation of D-RNA molecules in the cytoplasm of the cells. Cell division plays a causal role in determination since a slower rate of division allows the transcription *per cell* of more copies of the D-RNA being synthesized at that mitotic age, thus favoring accumulation and conservation of these D-RNA molecules. A more rapid rate of division gives transcription of fewer copies of D-RNA *per cell* characteristic of the mitotic age, thus effectively tending to bypass that period of competence. The number of cell divisions is important for determination since it allows the accumulation of more D-RNA, particularly the D-RNA present in few copies which may be transcribed from the less redundant genes, as well as being linked to a restriction in the number of kinds of D-RNA synthesized during this period of early development.

REFERENCES

Abbott, J., and Holtzer, H. (1968). The loss of phenotypic traits by differentiated cells. V. The effect of 5-bromodeoxyuridine on cloned chondrocytes. *Proc. Nat. Acad. Sci. U. S.* **59**, 1144–1151.

Balinsky, B. I. (1948). Korrelationen in der Entwicklung der Mund-und Keimenregion und des Darmkanals bei Amphibien. *Wilhelm Roux Arch. Entwicklungsmech. Organismen* **143**, 365–395.

Barth, L. G. (1966). The role of sodium chloride in sequential induction of the presumptive epidermis of *Rana pipiens* gastrulae. *Biol. Bull.* **131**, 415–426.

Barth, L. G., and Barth L. J. (1962). Further investigations of the differentiation *in vitro* of presumptive epidermis cells of the *Rana pipiens* gastrula. *J. Morphol.* **110**, 347–367.

Barth, L. G., and Barth L. J. (1963). The relation between intensity of inductor and type of cellular differentiation of *Rana pipiens* presumptive epidermis. *Biol. Bull* **124**, 125–140.

Brachet, J. (1965). The role of nucleic acids in morphogenesis. *Progr. Biophys. Mol. Biol.* **15**, 99–127.

Britten, R. J., and Kohne, D. E. (1965). Nucleotide sequence repetition in DNA. *Carnegie Inst. Washington Yearb.* **65**, 78–106.

Britten, R. J., and Kohne, D. E. (1968). Repeated sequences in DNA. *Science* **161**, 529–540.

Brown, D. D., and Littna, E. (1966). Synthesis and accumulation of DNA-like RNA during embryogenesis of *Xenopus laevis*. *J. Mol. Biol.* **20**, 81–94.

Brown, D. D., and Weber, C. S. (1968). Gene linkage by RNA-DNA hybridization. I. Unique DNA sequences homologous to 4S RNA, 5S RNA and ribosomal RNA. *J. Mol. Biol.* **34**, 661–680.

Clayton, R. M. (1968). A re-examination of the organ specificity of lens antigens. *Exp. Eye Res.* **7**, 11–29.

Daniel, J. C., and Flickinger, R. A. (in press). Nuclear DNA-like RNA in developing frog embryos. *Exp. Cell Res.*

Denis, H. (1966). Gene expression in amphibian development. II. Release of the genetic information in growing embryos. *J. Mol. Biol.* **22**, 285–304.

Flickinger, R. A. (1961). The effect of ethionine on the development of muscular response in the frog embryo. *J. Exp. Zool.* **147**, 21–32.

Flickinger, R. A. (1966). Reversible delay of normal development of frog embryos by inhibition of DNA synthesis. *J. Exp. Zool.* **161**, 243–250.

Flickinger, R. A. (1970). Effects of actinomycin D on RNA and protein synthesis in regions of developing frog embryos. *Experientia* **26**, 778–780.

Flickinger, R. A., and Coward, S. J. (1962). The induction of cephalic differentiation in regenerating *Dugesia dorotocephala* in the presence of the normal head and in unwounded tails. *Develop. Biol.* **5**, 179–204.

Flickinger, R. A., and Stone, G. (1960). Localization of lens antigens in developing frog embryos. *Exp. Cell Res.* **21**, 541–547.

Flickinger, R. A., Miyagi, M., Moser, C. R., and Rollins, E. (1967a). The relation of DNA synthesis to RNA synthesis in developing frog embryos. *Develop. Biol.* **15**, 414–431.

Flickinger, R. A., Freedman, M. L., and Stambrook, P. J. (1967b). Generation times and DNA replication patterns of cells of developing frog embryos. *Develop. Biol.* **16,** 457–473.

FLICKINGER, R. A., LAUTH, M. R., and STAMBROOK, P. J. (1970a). An inverse relation between the rate of cell division and RNA synthesis per cell in developing frog embryos. *J. Embryol Exp. Morphol.* **23,** 571–582.

FLICKINGER, R. A., DANIEL, J. C., and R. F. GREENE (1970b). The effect of rate of cell division on RNA synthesis in developing frog embryos. *Nature (London)* **228,** 557–559.

FLICKINGER, R. A., KOHL, D. M., LAUTH, M. R., and STAMBROOK, P. J. (1970c). Effect of rate and number of cell divisions on RNA synthesis. *Biochim. Biophys. Acta* **209,** 260–262.

GEBHARDT, D. O. E., and NIEUWKOOP, P. D. (1969). The influence of lithium upon the competence of the ectoderm in *Ambystoma punctatum. J. Embryol. Exp. Morphol.* **12,** 317–331.

GILLESPIE, D., and SPIEGELMAN, S. (1965). A quantitative assay for DNA-RNA hybrids with DNA immobilized on a membrane. *J. Mol. Biol.* **12,** 829–842.

GREEN, H., GOLDBERG, B., SCHWARTZ, M., and BROWN, D. D. (1968). The synthesis of collagen during the development of *Xenopus laevis. Develop. Biol.* **18,** 391–400.

GREENE, R. F., and FLICKINGER, R. A. (1970). Qualitative changes in DNA-like RNA during development in *Rana pipiens. Biochim. Biophys. Acta.* **217,** 447–460.

GROSS, J. (1963). Comparative biochemistry of collagen. *In* "Comparative Biochemistry" (M. Florkin and H. S. Mason, eds.), Vol. V, Part C, pp. 307–346. Academic Press, New York.

HOLTFRETER, J., and HAMBURGER, V. (1955). Amphibians. *In* "Analysis of Development (B. H. Willier, P. A. Weiss, and V. Hamburger, eds.), pp. 230–296. Saunders, Philadelphia, Pennsylvania.

HSU, T. C., SCHMID, W., and STUBBLEFIELD, E. (1964). DNA replication sequences in higher animals. *In* "The Role of Chromosomes in Development," (M. Locke, ed.), pp. 83–112. Academic Press, New York.

KAFIANI, C. A., TIMOFEEVA, M. J., MELNIKOVA, N. L., and NEYFAKH, A. A. (1968). Rate of RNA synthesis in haploid and diploid embryos of loach (*Misgurnus fossillis*). *Biochim. Biophys. Acta* **169,** 274–277.

KLOSE, J. and FLICKINGER, R. A. (in press). Collagen synthesis in frog embryo endoderm cells. *Biochim. Biophys. Acta.*

KOHL, D. M., GREENE, R. F., and FLICKINGER, R. A. (1969). The role of RNA polymerase in the control of RNA synthesis *in vitro* from *Rana pipiens* embryo chromatin. *Biochim. Biophys. Acta* **179,** 28–38.

KRUGELIS, E., NICHOLAS, J. S., and VOSGIAN, M. E. (1952). Alkaline phosphatase activity and nucleic acids during embryonic development of *Amblystoma punctatum* at different temperatures. *J. Exp. Zool.* **121,** 489–504.

LIEDKE, K. B. (1942). Lens competence in *Rana pipiens, J. Exp. Zool.* **90,** 331–347.

MASUI, Y. (1966). pH-dependance of the inducing activity of lithium ion. *J. Embryol. Exp. Morphol.* **15,** 371–386.

ÔGI, K. (1961). Vegetalization of the presumptive ectoderm of *Triturus* gastrula by exposure to lithium chloride solution. *Embryologia* **5,** 384–396.

RACHKUS, Y., TIMOFEEVA, M. Y., KUPRIYANOVA, N. S., and KAFIANI, C. (1969a). Transcription of repetitive DNA sequences in embryogenesis. *Mol. Biol. (USSR)* **3,** 338–347.

RACHKUS, Y., TIMOFEEVA, M. Y., KUPRIYANOVA, N. S., and KAFIANI, C. (1969b). Homologies in RNA populations at different stages of embryogenesis. *Mol. Biol. (USSR)* **3,** 486–493.

REMINGTON, J. A., and FLICKINGER, R. A. (in press). The time of DNA replication in the cell cycle in relation to RNA synthesis in frog embryos. *J. Cell. Physiol.*

SMITH, K. D., ARMSTRONG, J. L., and MCCARTHY, B. J. (1967). The introduction of radioisotopes into RNA by methylation *in vitro*. *Biochim. Biophys. Acta* **142**, 323–330.

STAMBROOK, P. J., and FLICKINGER, R. A. (1970). Changes in chromosomal DNA replication patterns in developing frog embryos. *J. Exp. Zool.* **174**, 101–114.

TAKATA, C., and YAMADA, T. (1960). Endodermal tissues developed from the isolated newt ectoderm under the influence of guinea pig bone marrow. *Embryologia* **5**, 8–20.

TEN CATE, G. (1953). "The Intrinsic Development of Amphibian Embryos." North Holland Publ., Amsterdam.

TWITTY, V. C. (1928). Experimental studies on the ciliary action of amphibian embryos. *J. Exp. Zool.* **50**, 319–344.

VILLEE, C. A. (1944). Phenogenetic studies of the homoeotic mutants of *Drosophila melanogaster*. II. The effects of temperature on proboscipedia. *J. Exp. Zool.* **96**, 85–102.

WHITELEY, A. H., MCCARTHY, B. J., and WHITELEY, H. R. (1966). Changing populations of messenger RNA during sea urchin development. *Proc. Nat. Acad. Sci. U. S.* **55**, 519–525.

WHITELEY, H. R., MCCARTHY, B. J., and WHITELEY, A. H. (1970). Conservatism of base sequences in RNA for early development of echinoderms. *Develop. Biol.* **21**, 216–242.

WOODLAND, H. R., and GURDON, J. B. (1968). The relative rates of synthesis of DNA, s-RNA and r-RNA in the endodermal region and other parts of *Xenopus laevis* embryos. *J. Embryol. Exp. Morphol.* **19**, 363–385.

Energy Metabolism in Early Mammalian Embryos[1]

THOMAS H. SHEPARD, TAKASHI TANIMURA,[2] AND
MAURICE A. ROBKIN

Central Laboratory for Human Embryology, Department of Pediatrics, and Department of Nuclear Engineering, University of Wasington, Seattle, Washington 98105

INTRODUCTION

Energy metabolism in the embryo has held the interest of biologists since the start of the scientific age. With the advent of isotope labeling techniques and the introduction of a superior method for *in vitro* culture of mammalian embryos (New, 1967), quantification of embryonic metabolism and some of the physiologic aspects of growth have become more attainable. This report deals with findings in normal somite stage rat embryos, and the data presented serve as control material for examining the mechanisms by which teratogenic agents may alter energy metabolism in the embryo. The intent of the paper is to summarize three aspects of energy metabolism carried out on rat embryos at somite stages: (a) direct observations of the embryos under different concentrations of oxygen, (b) utilization of glucose during whole embryo culture, and (c) studies of the activities of the succinic and DPNH oxidase systems (the terminal electron transport systems).

BIOLOGICAL MATERIAL

Breeding Methods

Female Sprague-Dawley rats, determined to be in estrus by the method of Blandau *et al.* (1941), were placed with individual males at about midnight. If vaginal sperma and copulation plugs were found the next morning (8:00 AM), pregnancy was diagnosed and that time was considered to be at the start of day zero (or first day).

Gestational Age, Somite Number, Crown-Rump Length, and Protein Content

Table 1 details the crown-rump length, somite number, and protein content of embryos after $10\frac{1}{3}$, 11, and 12 days of gestation. So-

---

[1] Supported by the National Institutes of Health grants HD02392, HD00180, HD00836, and by the Graduate School Research Fund, University of Washington.

[2] Present address: Department of Anatomy, Faculty of Medicine, Kyoto University, Sakyo-Ku, Kyoto, Japan.

TABLE 1
GROWTH MEASUREMENTS FOR SOMITE STAGE RATS

Age in days[a]	Crown-rump length (mm) (mean ± SE)	Somites (mean ± SE)	Protein (μg) (mean ± SE)
10⅓	2.0 ± 0.05	11.3 ± 0.3	45 ± 2
11	3.19 ± 0.05	25.3 ± 0.2	236 ± 5
12	5.46 ± 0.13	35.3 ± 0.5	730 ± 43

[a] The doe at estrus is placed with a male at about midnight, and, if vaginal sperm and copulation plugs are found, the next morning at 8:00 AM is considered the start of the first day and designated day zero. Day 10⅓ embryos were removed at 4:00 PM on the afternoon of the eleventh day (or day 10). The day 11 and day 12 embryos were removed between 9:00 and 10:00 AM on the two subsequent days.

mite number was estimated by counting in a caudal direction from somite 13, which was judged to be that somite located immediately caudal to the forelimb bud. The stage of limb bud development was recorded using the diagrams of the chick from Hamburger and Hamilton (1951). Protein was determined by the method of Lowry et al. (1951).

Blood Circulation during Rat Embryogenesis

Although heart contractions begin earlier, the rat embryo at the 10–12 somite stage has usually developed a regular heart beat. At approximately 17–19 somites, circulation within the yolk sac begins. At 24–26 somite stages considerable circulation is present in the yolk sac, but only beginning chorioallantoic circulation is present. At the 35-somite stage the chorioallanotic placental circulation is beginning to play a major role in comparison to the yolk sac circulation.

WHOLE EMBRYO IN VITRO STUDIES

Culture Method

The observations to be reported on heart rate and glucose utilization were obtained using the following *in vitro* method. Direct observation of the embryos during culture and collection of metabolites or intermittent addition of teratogenic substances is possible. Direct access to viable embryos enables study of physiologic and metabolic aspects of embryogenesis. Nicholas and Rudnick (1934) and Waddington and Waterman (1933) both realized the importance of *in vitro* observations of growing mammalian embryos. A number of similarities exist between their techniques and the modification of the procedure described by New (1967).

Readers are referred to New's description, which is more detailed than that given here (New and Stein, 1964; New, 1967). Another support system for mammalian embryos has been described by Tamarin and Jones (1968). Both methods take advantage of the well developed yolk sac placenta in the rat and mouse.

After sacrifice by cervical dislocation of the maternal rat on day $10\frac{1}{3}$, 11, or 12, the pear-shaped decidual masses are separated from the uterus. The second step is removal of the decidua from the embryonic tissue; this is performed by gentle opposing traction with two jeweler's forceps observed under 10–15× magnification. The resulting embryonic tissue consists of an inverted yolk sac covered to a varying extent by chorioallantoic placenta. The next step, technically the most difficult, is opening of the chorioallantoic covering along with Reichert's membrane and the parietal yolk sac. The same forceps are used, but a high magnification is helpful. The explant is then attached to a rayon acetate raft by pressing Reichert's membrane firmly to the surface (Fig. 1). The rafts are previously prepared by immersion in a solution of rat tail collagen and sterilized; they are then weighted by attachment to a glass coverslip. The application of Reichert's membrane to the raft is simplified by manipulation with two smooth-tipped glass rods. All the above procedures are carried out beneath the surface of a balanced salt solution using varying magnifications of the dissecting microscope. Good jeweler's forceps are essential. The next step is transfer of the rafts, while still immersed, into one of New's glass circulators (Fig. 2). The circulator is gassed by bubbles which pass up the side arm producing circula-

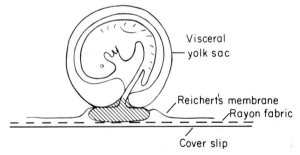

FIG. 1. Diagram to show attachment of embryo and placenta to a raft in preparation for explanation.

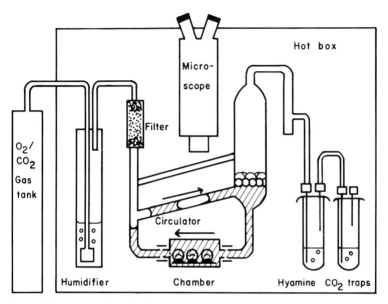

Fig. 2. Diagram to show *in vitro* culture of whole embryos. Gas is humidified and passed through a rouge filter; then by propelling bubbles up the side arm of New's glass circulator, flow of the aerated serum is established. The outflow of CO_2 may be trapped, if indicated, in two test tubes containing hydroxide of Hyamine. The equipment and dissecting microscope are contained in a hot box constructed of Lucite.

tion of the medium. The culture medium consists of either rat or human serum; we find human serum to give results similar to those of the rat. The entire procedure is carried out using sterile precautions.

The circulator is maintained in a warming box at 37°–38°C. For the 10-day embryos a mixture of air and 5% CO_2 is used; for the older preparations 95% O_2 and 5% CO_2 is necessary (see below for further discussion). Direct observations of the embryos in the chamber through the yolk sac is possible and the yolk sac circulation itself can also be well seen.

Growth of 11-Day, 25-Somite Embryo in Vitro

During the first 3 hours after explantation on day 11 a vigorous heart action and yolk sac circulation is found. During this 3-hour explant period there is an increase of 1.4 ± 0.3 somites. After 3–5 hours there is a gradual decline in heart rate and in the day-11 embryos beginning sludging of the yolk sac circulation begins. When day-11 embryos are explanted and examined after about 22 hours of culture, the somite number has increased from 25 to 31 or 32 and the protein content has about doubled. These 22-hour results are less than shown for 12 days of *in vivo* growth (Table 1). Growth as measured by protein increment is proportionally less than somite addition in this "long term" *in vitro* system.

A number of problems with the method still exist. Establishment of full yolk sac circulation between day 10 and day 11 seldomly occurs; New has observed that the appearance of an intraembryonic blood bypass around the placental circulation is a feature of this problem. Toward the middle part of day 12 the oxygen requirements of the embryo cannot be fully met by the circulator. New has ameliorated this to some extent by maintaining the preparation in hyperbaric oxygen (New and Coppola, 1970).

HEART RATE CHANGES AT DIFFERENT OXYGEN CONCENTRATIONS

Rat embryos were explanted at the three somite stages described in Table 1, and the heart rates were counted directly during a 2-hour period while the gas delivered to the chamber varied between 95% oxygen and CO_2, air and CO_2, and nitrogen (Shepard et al., 1969). Oxygen levels were monitored by a Clark electrode inserted into the chamber. The initial heart rates of embryos of these $10\frac{1}{3}$-, 11-, and 12-day stages were found shortly after explantation of average 154–163 (Table 2).

Emergence of Oxygen Dependence after Day $10\frac{1}{3}$

The day-$10\frac{1}{3}$ embryos after exposure to air or nitrogen evidence a small and statistically nonsignificant decrease in heart rate (Table 2). It was possible to maintain day-$10\frac{1}{3}$ embryos in nitrogen for up to 30 minutes with little effect on heart rate. Significant drops in heart rate of the day 11 and day 12 embryos were observed 10 minutes after exposure to air or nitrogen. After longer periods of nitrogen the 11- and 12-day stages developed decreased amplitude of the heart beat and then sludging in their yolk sac circulation. The yolk sac blood stasis became irreversible and contributed to eventual embryonic death.

Oxygen Toxicity of Day-$10\frac{1}{3}$ Embryos

Day-$10\frac{1}{3}$ embryos developed poorly if exposed to 95% oxygen for any length of time. Within an hour of high oxygen exposure the

TABLE 2
INITIAL HEART RATES AND CHANGES IN RATE AFTER GAS EXPOSURE[a]

(1) Day	(2) Initial rate (Mean ± SE)	No.	(3) Decrease after 10 min in air (Mean ± SE)	No.	(4) Decrease after 10 min in nitrogen (Mean ± SE)	No.
(a) 10⅓	154 ± 7	13	12 ± 5	13	22 ± 7	11
(b) 11	162 ± 2	46	33 ± 8	8	50 ± 6	12
(c) 12	163 ± 4	15	41 ± 6	14	62 ± 8	13

[a] From Shepard et al. (1969).

[b] Statistical analysis using the probabilities from the table of distribution of t of Fisher and Yates (1963) indicated $p < 0.001$ for 3a vs 3c, 4a vs 4c, 2b vs 3b, 2c vs 3c; $p < 0.01$ for 4a vs 4b; $p < 0.05$ for 3a vs 3b; $p > 0.05$ for 2a vs 2b, 3b vs 3c, 4b vs 4c, 2a vs 3a, and 2a vs 4a. Where different columns were compared, the actual rates in columns 3 and 4 were used.

heart rate began to decrease, and within a 2- to 3-hour period all heart beats had stopped. This sensitivity to oxygen was completely lost by day-11 embryos; they required high concentrations of oxygen for survival. We have no adequate explanation for this phenomenon. The observations by New and Stein (1964) that exposure to air produced better growth than 60% oxygen in presomite embryos is consonant with this observed oxygen damage.

GLUCOSE METABOLISM

Methods of Study

Rat embryos at the three stages of somite development outlined in Table 1 were explanted into the *in vitro* system described above. All preparations were maintained in a viable state which was confirmed by visual observations of the heart rate and yolk sac circulation. Uniformly labeled ^{14}C-glucose was added to the medium in a final concentration of 0.25–1.25 μCi per milliliter of medium. The total volume of the serum medium was 10 ml, and three embryos were explanted in each circulator. For 3 hours after addition of the isotope the metabolic fate of the glucose was measured by determining liberated $^{14}CO_2$, ^{14}C-lactate production in the medium, and ^{14}C incorporation into the embryo and membranes. The $^{14}CO_2$ was trapped in the outflow gas in hydroxide of Hyamine with the use of successive traps in order to ascertain completeness of the trapping method. By the method of Cuppy and Crevasse (1963), $^{14}CO_2$ was liberated from the medium and lactate in the serum was separated by the method of Barker and Summerson (1941). A thin-layer chromatography technique was used to check the purity of the added isotopes and to ascertain that ^{14}C lactate was the only measurable labeled organic acid product in the serum. The glucose level in the serum was determined by the glucose oxidase method (Marks, 1959).

For calculations of glucose utilization the sum of the ^{14}C counts from the tissues, CO_2, and lactate was expressed as "utilized." The amount of radioactivity added to the serum at the start represented 100%. On the assumption that cold glucose is utilized in the same manner as labeled glucose, the glucose utilization (μmoles of glucose per gram of protein per hour, Gu) was calculated by the following formula: Gu = $(U \times g \times 55,600)/P$ where U is percentage of ^{14}C utilized; g is mg of glucose in the medium; 55,600 is a constant for conversion to μmoles; and P is total protein content of embryos and membranes in micrograms.

High Utilization at Day 10⅓

The average calculated utilization rate of glucose at 10⅓ days was 731 μmoles per gram of protein per hour. On day 11 this rate dropped to 474 μmoles per gram of protein per hour; and on day 12, to 312 μmoles per gram per hour (Table 3, Fig. 3) (Tanimura and

TABLE 3
Glucose Utilization by Rat Embryos Cultured for 3 Hours
(Experiments with U-^{14}C-Glucose)

Day of gestation	Number of embryos	μMoles glucose/gm protein/hour (mean ± SE)[a]			
		Total calculated utilization	Embryos + membrane (%)[b]	CO_2 (%)[b]	Lactate (%)[b]
10⅓	15	731 ± 91	39 ± 4 (5.6)	33 ± 4 (4.6)	659 ± 88 (89.8)
		∨∨	∧	∥	∨∨
11	24	474 ± 32	54 ± 2 (11.6)	40 ± 2 (8.6)	381 ± 30 (79.8)
		∨	∧	∥	∨∨
12	8	312 ± 39	66 ± 5 (21.7)	34 ± 6 (10.7)	212 ± 30 (67.6)

[a] ∥: $P > 0.05$, ∨: $P < 0.05$, ∨∨: $P < 0.01$.
[b] Percentage of the total calculated utilization.

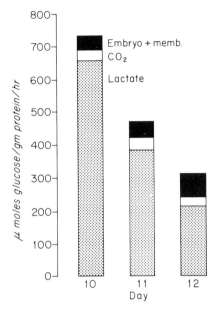

FIG. 3. *In vitro* utilization of glucose by rat embryos calculated from studies using uniformly labeled glucose-^{14}C. Note sharp drop in lactate production after day 10.

Shepard, 1970). A slight increase in the number of micromoles of glucose incorporated into the embryo and membranes was noted with increasing maturation (Table 3). The extensive literature on the general subject of glucose metabolism has been examined with special attention to studies of intact systems that give the highest rate of glucose uptake, the object being to compare the glucose metabolism of living embryos with that of other tissues. To arrive at some common denominator to enable comparison, the data reported earlier, originally given per unit wet weight, were converted to micromoles of glucose per gram of protein per hour based on the best estimate of protein content in the corresponding tissue (Long, 1961). This conversion gave the following values in these tissues of the rat: perfused heart, 267 and 515; perfused liver, 118; diaphragm, 66; brain slices, 107; lens, 29. A comparison of the glucose utilization of these tissues with those in Table 3 leads to the conclusion that $10\frac{1}{3}$-day rat embryos metabolize glucose at a high rate, higher than that of perfused heart tissue (Vahouny *et al.*, 1966).

A few studies, in addition to ours, are available on embryonic tissues. DePlaen (1969) reported that isolated rat embryos used glu-

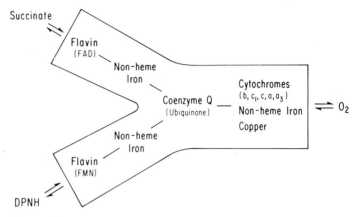

Fig. 4. Schematic composition of the electron transport particle (ETP).

cose at a rate of 302 μmoles/gram protein/hour when cultured in Warburg flasks for 3 hours. This value dropped to 23 for older embryos; both figures are significantly lower than ours. Probably embryos being in Warburg flasks as opposed to growing in culture accounts for the decreased metabolism he found. Studies by Fridhandler et al. (1967) showed that glucose utilization by day 6 rabbit preimplantation blastocysts is 410 μmoles/gram protein/hour.

Incorporation and CO_2 and Lactate Production

The drop in glucose utilization rate between day $10\frac{1}{3}$ and day 12 was associated with a sharp decline in the proportion of lactate produced (Table 3, Fig. 3). The lactate produced at $10\frac{1}{3}$ days' gestation was 659; and at 12 days, 212 μmoles per gram per hour. The micromoles of CO_2 produced per gram of protein per hour from glucose did not change from day $10\frac{1}{3}$ to day 12. A slight increase in the number of micromoles of glucose incorporated per gram of protein in the embryos and membranes was noted with increasing maturation (Table 3).

Pentose Phosphate Pathway

Attempts were made to determine the role of the pentose phosphate pathway for glucose utilization by comparing $^{14}CO_2$ production during a 3-hour period from glucose labeled on the first carbon with that labeled on the sixth carbon (Tanimura and Shepard, 1970). In most mature tissues the C1:C6 ratio is usually about 1, indicating that almost all the glucose is metabolized via the Embden-Meyerhof pathway. Noted exceptions may be adipose tissue, mammary gland, and some endocrine glands. The ratio of $^{14}CO_2$ from C1- to C6-labeled glucose was 11.3 on day $10\frac{1}{3}$, 3.6 on day 11, and 4.6 on day 12. Carbon 1 of glucose was more rapidly metabolized into $^{14}CO_2$ than C6 suggesting that relatively more glucose was converted to pentose during the earlier stage of rat embryonic development.

In embryonic tissues it is generally assumed that the pentose phosphate pathway has a more important role in glucose metabolism as compared with adult tissues. This pathway is considered to be important for synthesis of ribose precursors of nucleic acids and for reduction of nicotinamide adenine dinucleotide phosphate.

Most data on mammals so far reported are based on experiments with fetuses in later gestation rather than young embryos and show a C1:C6 ratio much lower than found here (Villee and Loring, 1961). Coffey et al. (1964) reported that the C1:C6 ratio of chick embryo heart homogenates was 5 on days 2–4 and fell sharply to almost 1 at a later stage. DeMeyer and DePlaen (1964) noted that the C1:C6 ratio in rat embryos in Warburg flasks incubated for 30 minutes was 28 and stressed the relative importance to the embryo of the pentose phosphate pathway. Unlike the somite stages of embryogenesis, studies on the C1:C6 ratio of $^{14}CO_2$ yield by preimplantation ova by Fridhandler (1961) in rabbits, Brinster (1967) in mice, and Brinster (1968) in rabbits showed a relatively low ratio (1.1–1.4) at the blastocyst stage when incubated 1–4 hours.

TABLE 4
TERMINAL ELECTRON TRANSPORT ACTIVITY IN RAT EMBRYOS OF DAY 10 TO DAY 13

Day of gestation	Average specific activity[a] in embryo		
	DPNH → O_2 plus cytochrome c		Succinate → O_2 cytochrome c
	Antimycin[b] sensitive	Antimycin insensitive	
10	0.0055	0.0103	—
11	0.0101	0.0200	—
12	0.0169	0.0141	0.0113
13	0.0180	0.0111	0.0142

[a] Micromoles of substrate oxidized per minute per milligram of protein.
[b] Concentration of antimycin (1 µg/ml) produced maximal inhibition in the assays. Each value is based on the average of at least five experiments, each of which included about 40 embryos.

Comparison of in Vitro and in Vivo Glucose Utilization

Validity of the *in vitro* studies were tested by *in vivo* studies of the metabolism of glucose-^{14}C. Pregnant rats were given intravenously 100 μCi of uniformly labeled glucose on day 11, and 3 hours later the embryos were removed and the amount and distribution of the ^{14}C was studied. The counts per minute per milligram of protein were 31,600 in the embryo, 16,000 in the membranes, 5350 in the decidua, but only 3300 in the maternal liver. This suggests that the rat embryo, at this stage, does have the ability under normal physiologic states to concentrate high amounts of glucose from the mother's circulation. The chemical distribution of the ^{14}C in embryos from the *in vivo* experiment was found to be roughly similar to that found after the *in vitro* studies (Tanimura and Shepard, unpublished). In summary, the *in vivo* studies showed distribution of embryonic ^{14}C in three fractions: (1) trichloroacetic acid soluble, (2) lipid, and (3) protein plus nucleic acids was 32, 21, and 48%, respectively, whereas the embryos after *in vitro* culture had 54, 13, and 33%, respectively. The proportional increase in lipid and protein nucleic acid fraction incorporation *in vivo* may be in part explained by the conversion of glucose to fatty and amino acids by the maternal tissues.

TERMINAL ELECTRON TRANSPORT SYSTEM

The terminal electron transport system is an important enzyme responsible for generation of cellular energy from molecular oxygen and inorganic phosphate. The electron transport system, which is a large particulate enzyme of molecular weight of approximately two million, carries out both the reoxidation of reduced diphosphopyridine nucleotide or DPNH and the oxidation of succinate, with concomitant formation of energy by the process termed oxidative phosphorylation. The metabolic products of glycolysis are oxidized via the Krebs or citric acid cycle. There are five sites of oxidation, four of which are mediated by the reduction of diphosphopyridine nucleotide and one by the direct oxidation of succinate. The enzyme system is located in the mitochondria and, when the cells and mitochondria are fragmented, the electron transport particle or ETP can be isolated as a purified enzyme. The electron transport particle of mammalian cells is composed of several active components or coenzymes which are arranged schematically as shown in Fig. 4, and which consists of the various cytochromes, iron, copper, ubiquinone, and, in particular, two flavin groups: flavin mononucleotide, or FMN, and flavin adenine dinucleotide, or FAD.

Assay Method

The enzyme assays were performed on embryos from Sprague-Dawley rats maintained and bred in the same manner as those used in the experiments above, and the biochemical determinations were carried out under the supervision of Dr. Bruce Mackler (Aksu et al., 1968). Assays of homogenates of pooled embryos are expressed in Table 4 as micromoles of substrate oxidized per milligram of total embryonic protein. For embryos of day 12 or older it was possible to confirm the homogenate findings by studies of a partially purified electron transport particle from differential centrifugation. Since all tissue fractions were at best only partially purified preparations of the electron transport system, they also contained other enzymatic systems capable of oxidizing DPNH. However, the activity of the electron transport systems can be separated from that of other DPNH-oxidizing systems, since only the former activity is inhibited by an antibiotic called antimycin A. Thus, DPNH oxidase activity was determined as antimycin-sensitive activity (activity of the electron transport system) and antimycin-insensitive activity. It was necessary to determine both succinate and DPNH oxidase activities in the presence of externally added cytochrome c, since cytochrome c was partially removed from preparations of the electron transport

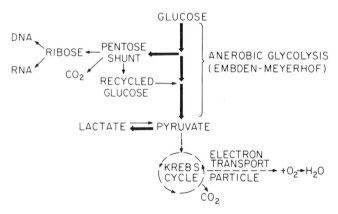

FIG. 5. Simplified diagram of energy metabolism from carbohydrate. The heavy lines represent active pathways around the 11-somite stage and the dashed lines show emergence of the Krebs cycle–electron transport system in the later somite embryos. (From Mackler, 1969.)

systems during purification procedures, and the addition of cytochrome c was necessary for maximal enzymatic activity.

Increase from Day 10 to Day 12

The activity of the antimycin-sensitive DPNH portion of the electron transport system increases 3-fold between day 10 and day 12 (Table 4). Activity of the succinoxidase system is attaining a plateau between day 12 and day 13. We were unable to muster enough day-10 and -11 embryos to accurately measure oxidation of succinate. These studies (Aksu et al., 1968; Mackler, 1969) have served as controls for experiments demonstrating that the electron transport system was specifically decreased during the genesis of the syndrome of congenital anomalies produced by riboflavin deficiency in the rat.

INTERCORRELATIONS OF OXYGEN REQUIREMENT, GLUCOSE METABOLISM, AND THE TERMINAL ELECTRON TRANSPORT SYSTEM

A diagrammatic representation of the various pathways under discussion in this paper is given in Fig. 5. The glycolytic pathway of Embden-Meyerhof operates anerobically while the pentose phosphate shunt and the Krebs citric acid cycle and terminal electron transport system require oxygen. The pentose shunt even at maximal activity oxidizes considerably less carbon than the Krebs cycle–electron transport system. In the above studies we have measured input of glucose and the output of CO_2 and lactate, and through biochemical means we have studied the ability of the terminal electron transport system to utilize ozygen. Therefore, it is possible to make at least some general statements about what sort of reactions are occurring in the system diagrammed in Fig. 5.

Before proceeding further we wish to point out that we do not have information on the contributions of fatty and amino acids to this metabolic system; accordingly, we are describing mainly energy metabolism in terms of carbohydrate utilization. Another slightly tenuous assumption is, of course, that the *in vitro* systems we have used do actually maintain the embryos in a physiologically normal growing state. Our *in vivo* studies at least confirm high rates of glycolysis in the day-11 rat embryos, but the proportion of carbon entering the lipid and nitrogenous fractions is somewhat more than is found in embryos grown *in vitro*. In working with mammalian embryos, we are limited in the amount of tissue that can be obtained for analysis; fortunately, portions of these metabolic systems have been previously explored in amphibian, avian, and sea urchin embryos, and in many respects the biochemical mechanisms of different species seem to be similar.

In the early somite stage embryos (day $10\frac{1}{3}$; 11 somites) we have considerable evidence that glycolysis is actively proceeding through the Embden-Meyerhof pathway but that products of this pathway are not being oxidized through the Krebs cycle–electron transport system pathway. The evidence is (a) a very high glucose utilization rate with high amounts catabolized lactate appearing, (b) a relative lack of oxygen dependence as judged by direct observation of heart rate, and (c) measurably low activity of terminal electron transport system. During this stage of development effective heart action and early intraembryonic circulation is only beginning to be established; the yolk sac and chorioallantoic circulations do not exist. One might surmise that relatively anerobic conditions exist within the embryo.

During subsequent somite stages (25 and 35 somites) in the rat embryo there occurs a gradual emergence of oxygen dependence and increase in terminal electron transport system activity (designated by dotted lines in Fig. 5). This metabolic shift is associated with establishment of the yolk sac and then the chorioallantoic circulation. Considerable increase in the number of erythrocytes is taking place during this transitional phase.

The pentose phosphate shunt pathway plays a significant role in glucose metabolism during these somite stages, and a decrease in this activity occurs with advancing age. An obvious role of this ribose-producing pathway exists in cells synthesizing nucleic acids. The CO_2 ratio method required by the small sample sizes in our study of the role of the pentose phosphate pathway has been shown to give only relative values; more precision can be obtained by the use of specific activities of the glycolytic products (Katz and Wood, 1963).

Although it is not the purpose of this paper to review the extensive literature on oxidative metabolism in other species of embryos, evidence is available to support the contention that the general pattern of early energy metabolism is similar in most embryonic species. Spratt (1958) in classic experiments demonstrated the ability of glucose to maintain early embryonic chick embryos *in vitro*. Subsequent reviews by Herrmann and Tootle (1964) and by Romanoff (1967) provide details about oxidative metabolism in other embryonic species. The work of Sippel (1954) on succinoxidase activity of heart ventricles from somite stage chick and rat embryos shows a steep increment in activity during the same developmental periods described in this paper. He found a 3-fold increment in succinoxidase activity during the same developmental period where we show an increase in antimycin-sensitive DPNH oxidation.

Changes in the number of mitochondria and their internal morphology have been reviewed by Herrmann and Tootle (1964) and by Gustafson (1965). An increase in number of mitochondria and in the number of their cristae appears to occur during the somite stages comparable to those described in this paper.

SUMMARY

Three somite stages of the rat embryos used (11, 25, and 35 somites) in these experiments have been characterized by gestational period, morphology, and protein content. The *in vitro* method used for maintaining these three stages of embryos in a growing state while they were studied is described.

The earliest stage studied (about 11 somites) is characterized by relatively high glucose utilization rate with the major end product appearing as lactate. Furthermore, this early stage when exposed to lowered oxygen evidenced much less fall in heart rate than the 25- and 35-somite stages. The specific activity of the electron transport system as measured in homogenates from the 11-somite embryos is only one-third that seen in the 35-somite embryos. These three pieces of evidence suggest that early embryonic metabolism is characterized by high rates of anaerobic glycolysis (Embden-Meyerhof pathway) with relatively little activity of the Krebs cycle–electron transport system.

By the 35-somite stage the rate of glucose utilization dropped along with a decrease to about 30% in lactate production, oxygen requirement appeared, and a 3-fold increase in antimycin-sensitive DPNH oxidase appeared. These changes indicate increased activity of the Krebs cycle–electron transport system energy pathway during development from the 11- to the 35-somite stage.

The pentose pathway was studied using the ratio of CO_2 derived from C1 to C6 labeled glucose. The results showed that at the 11-somite stage a relatively high proportion of glucose was being utilized by the pentose phosphate pathway and that this proportion decreased at later stages.

The 11-somite embryos are damaged by high concentrations of oxygen.

REFERENCES

AKSU, O., MACKLER, B., SHEPARD, T. H., and LEMIRE, R. J. (1968). Studies of the development of congenital anomalies in embryos of riboflavin-deficient, galactoflavin fed rats. II. Role of the terminal electron transport systems. *Teratology* 1, 93–102.
BARKER, S. B., and SUMMERSON, W. H. (1941). The colorimetric determination of lactic acid in biological material. *J. Biol. Chem.* 138, 535–554.
BLANDAU, R. J., BOLING, J. L., and YOUNG, W. C. (1941). The length of heat in the albino rat as determined by a copulatory response. *Anat. Rec.* 79, 453–463.
BRINSTER, R. L. (1967). Carbon dioxide production from glucose by the preimplantation mouse embryo. *Exp. Cell Res.* 47, 271–277.
BRINSTER, R. L. (1968). Carbon dioxide production from glucose by the preimplantation rabbit embryo. *Exp. Cell Res.* 51, 330–334.
COFFEY, R. G., CHELDELIN, V. H., and NEWBURGH, R. W. (1964). Glucose utilization by chick embryo homogenates. *J. Gen. Physiol.* 48, 105–112.
CUPPY, D., and CREVASSE, L. (1963). An assembly for $C^{14}O_2$ collection in metabolic studies for liquid scintillation counting. *Anal. Biochem.* 5, 462–463.
DEMEYER, R., and DEPLAEN, J. (1964). An approach to the biochemical study of teratogenic substances on isolated rat embryo. *Life Sci.* 3, 709–713.
DEPLAEN, J. L. (1969). Aspects dynamiques du métabolisme glucidique chez l'embryon de rat. Action de substances hypoglycémiantes et tératogènes. Thèses. Edition Arscia S.A., Bruxelles.
FISHER, R. A., and YATES, F. (1963). "Statistical Tables." Hafner, New York.
FRIDHANDLER, L. (1961). Pathways of glucose metabolism in fertilized rabbit ova at various pre-implantation stages. *Exp. Cell Res.* 22, 303–316.
FRIDHANDLER, L., WASTILA, W. B., and PALMER, W. M. (1967). The role of glucose in metabolism of the developing mammalian preimplantation conceptus. *Fertil. Steril.* 18, 819–830.
GUSTAFSON, T. (1965). Morphogenetic significance of biochemical patterns in sea urchin embryos. In "The Biochemistry of Animal Development," (R. Weber, ed.). Academic Press, New York.
HAMBURGER, V., and HAMILTON, H. L. (1951). A series of normal stages in the development of the chick embryo. *J. Morphol.* 88, 49–92.
HERRMANN, H., and TOOTLE, M. L. (1964). Specific and general aspects of the development of enzymes and metabolic pathways. *Physiol. Rev.* 44, 289–371.
KATZ, J., and WOOD, H. G. (1963). The use of $C^{14}O_2$ yields from glucose-1 and glucose 6-C^{14} for the evaluation of the pathways of glucose metabolism. *J. Biol. Chem.* 238, 517–523.
LONG, C. (1961). "Biochemist's Handbook." Van Nostrand, Princeton.
LOWRY, O. H., ROSEBROUGH, N. J., FARR, A. L., and RANDALL, R. J. (1951). Protein measurement with Folin phenol reagent. *J. Biol. Chem.* 193, 265–275.
MACKLER, B. (1969). Studies of the molecular basis of congenital malformations. *Pediatrics* 43, 915–926.
MARKS, V. (1959). An improved glucose-oxidase method for determining blood, C.S.F., and urine glucose levels. *Clin. Chim. Acta* 4, 395–400.
NEW, D. A. T. (1967). Development of explanted rat embryos in circulating medium. *J. Embryol. Exp. Morphol.* 17, 513–525.
NEW, D. A. T., and COPPOLA, P. T. (1970). Development of explanted rat fetuses in hyperbaric oxygen. *Teratology* 3, 153–159.

New, D. A. T., and Stein, K. F. (1964). Cultivation of postimplantation mouse and rat embryos on plasma clots. *J. Embryol. Exp. Morphol.* **12,** 101–111.

Nicholas, J. S., and Rudnick, D. (1934). Development of rat embryos in tissue culture. *Proc. Nat. Acad. Sci. U.S.* **20,** 656–658.

Romanoff, A. L. (1967). "Biochemistry of the Avian Embryo." Wiley, New York.

Shepard, T. H., Tanimura, T., and Robkin, M. A. (1969). In vitro study of rat embryos I. Effects of decreased oxygen on embryonic heart rate. *Teratology* **2,** 107–109.

Sippel, T. O. (1954). The growth of succinoxidase activity in the hearts of rat and chick embryos. *J. Exp. Zool.* **126,** 205–221.

Spratt, N. T. (1958). Chemical control of development. *In* "The Chemical Basis of Development" (W. D. McElroy and B. Glass, eds.). Johns Hopkins Press, Baltimore, Maryland.

Tamarin, A., and Jones, K. W. (1968). A circulating medium system permitting manipulation during culture of postimplantation embryos. *Acta Embryol. Morphol. Exp.* **10,** 288–301.

Tanimura, T., and Shepard, T. H. (1970). Glucose metabolism by rat embryos in vitro. *Proc. Soc. Exp. Biol. Med.* **135,** 51–54.

Vahouny, G. V., Katzen, R., and Entenman, C. (1966). Glucose uptake by isolated perfused hearts from fed and fasted rats. *Proc. Soc. Exp. Biol. Med.* **121,** 923–928.

Villee, C. A., and Loring, J. M. (1961). Alternative pathways of carbohydrate metabolism in foetal and adult tissues. *Biochem. J.* **81,** 488–494.

Waddington, C. H., and Waterman, A. J. (1933). The development in vitro of young rabbit embryos. *J. Anat.* **67,** 355–370.

Control of Differentiation in *Volvox*

RICHARD C. STARR

Department of Botany, Indiana University, Bloomington, Indiana 47401

INTRODUCTION

Progress in understanding a biological process and its control is dependent upon, and at the same time restricted by, the nature of the organism which provides the system under investigation. Tremendous advances have been made toward our understanding of development on the molecular and cellular level through such organisms as viruses, bacteria, the ciliated protozoa, and the alga *Acetabularia*, but many questions concerning development in multicellular organisms will probably never be answered using the prokaryotes and the unicellular microorganisms, or the very complex systems now available in vertebrate animals or the seed plants. The need is great, accordingly, for organisms more auspicious for the investigation of development.

The green flagellate *Volvox* provides us with a system in which to study differentiation in a simple multicellular organism having only two kinds of cells, somatic and reproductive. The somatic cells are always the same in structure and function, but the reproductive cells may differ in number, origin, position, and function depending on whether the individual is asexual, male, or female. *Volvox* is a haploid, autotrophic microorganism capable of extended multiplication in the axenic state through asexual reproduction in a simple, chemically defined medium. Through the use of a species specific inducer, the development of male or female individuals, rather than asexual ones, can be effected.

Over 60 years ago Powers (1907, 1908) was impressed with *Volvox* as promising material for developmental studies, and as recently as 1964 Barth pointed out that the family of colonial flagellates to which *Volvox* belongs presents "the problem of the fundamental control of somatic versus germ cell determination in such challenging simplicity that one wonders why contemporary investigators do not turn to them as materials for fresh inroads into the problem."

In 1963 Dr. William Darden, at that time a graduate student in my laboratory, began a study of *Volvox aureus* Ehrbg. in an attempt to see whether the phenomenon of sexual isolation could be demonstrated in this genus as it had been in other colonial green flagellates, *Gonium* (Stein, 1958), *Pandorina* (Coleman, 1959), and *Eudorina* (including *Pleodorina*) (Goldstein, 1964). Darden's observations that sexual stages could be evoked in *Volvox aureus* in a medium designed by Provasoli and Pintner (1959) and his subsequent research into the control of the sexual process (Darden 1965, 1966) initiated our interest in the use of the genus for the study of development.

Biologists have been fascinated with *Volvox* since the first record of its existence by Leeuwenhoek in a letter dated January 2, 1700, describing observations he had made in August and September, 1698 (Dobell, 1932). In the ensuing years it has been studied by botanists, zoologists, phycologists, and protozoologists, these studies resulting in a plethora of terms, at times synonymous, conflicting or contradictory. Therefore, in speaking about *Volvox* as a multicellular organism to an audience in which many may consider it as a colonial protozoan while others are convinced it is a colonial alga, I must first offer this apologia by Powers (1907) from his paper "New Forms of *Volvox*":

> "I shall deliberately use, in this paper, such expressions as 'somatic cells,' 'reproductive cells,' etc., and I shall also use, with equal freedom, the terms 'colony,' 'coenobium,' and the like. Such expressions are flatly contradictory, in a sense; but so are the facts. They confuse no one familiar with the different points of view from which *Volvox* may be considered. The very beauty of *Volvox* and its group lies in the happy way in which they override 'fundamental distinctions.' Out of the seeming chaos, therefore, of terms old and terms new, terms botanical and terms zoological, terms metazoon and terms protozoon, I choose those most convenient and useful for the context in question."

When Darden was completing his study on *Volvox aureus* in 1965, Dr. Gary Kochert, then a graduate student at Indiana University, began his work with another species, *Volvox carteri* f. *weismannia* (Powers) Iyengar. In this species Kochert (1968) described a system in which the female strain required an inducing factor produced by sexual males in order to differentiate eggs rather than asexual reproductive cells. He was not able to demonstrate any susceptibility of the male strain to its own extract, relying on "spontaneous" production of males following a transfer to fresh medium.

In March, 1967, male and female strains of *V. carteri* f. *nagariensis* Iyengar were isolated from soil collections made near the edge of a small pond, a segment of a network of ponds and ditches in the paddy fields surrounding Befu near Kobe, Japan. Unlike the strains of *V. carteri* f. *weismannia* used by Kochert (1968), the differentiation of both male and female individuals could be effected through the use of an inducer in the fluid from sexual male cultures, thereby providing two sexual systems for analysis and also allowing for the controlled production of large quantities of males required for the subsequent isolation of the chemical inducer. It is with this pair of strains (HK 9 and 10), and strains derived from them by spontaneous mutation and/or recombination, that I have been concerned for the past three years and discuss in detail herein.

VOLVOX CARTERI f. NAGARIENSIS IYENGAR

Cultivation of the male and female strains presents no particular problems. The methods for growing and maintaining the asexual phase have been presented elsewhere (Starr, 1969).

The asexual individual of *Volvox carteri* f. *nagariensis* is composed of two kinds of cells, somatic and reproductive, arranged on the periphery of a spheroid. There may be as many as 5000 small somatic cells, but there are rarely more than 16 large asexual reproductive cells which are located in the posterior region of the spheroid (Figs. 1–3). The biflagellate protoplasts of the somatic cells are surrounded by a gelatinous cellular envelope, but there are no cytoplasmic connections between adjacent protoplasts in this or other species belonging to the taxonomic Section Merrillosphaera. Cytoplasmic connections are found only in the other three sections of the genus (Smith, 1944).

Sexual spheroids are formed under the influence of an inducer produced by sexual males. The female strain forms female spheroids bearing 35–45 eggs (Fig. 4) in contrast to the 16 gonidia found in the asexual spheroid (Fig. 3) of the same strain. The male strain forms small male spheroids of reduced cell number; these dwarf males have a 1:1 ratio of somatic cells to reproductive cells, the latter the androgonidia (Fig. 5), which undergo division to form packets of 64 or 128 sperm (Fig. 6). The asexual spheroids in the male strain are identical to those in the female strain.

Embryogenesis

The development of all types of spheroids—asexual, male, and female—is through successive divisions of the asexual reproductive cells, the gonidia. In both the male and female strains, the increase in size of a population is effected by the production of asexual spheroids by gonidia, the process being repeated in successive generations. However, if an inducing substance produced in sexual male

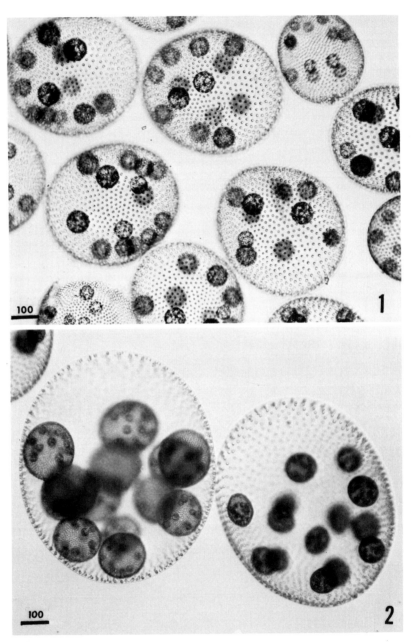

Unless otherwise noted, all figures are of *Volvox carteri* f. *nagariensis*. The number appearing above the marker on each figure indicates the scale in microns.

Fig. 1. Asexual spheroids with enlarging gonidia.

Fig. 2. Parental spheroids in which new asexual individuals have been formed by cleavage of the gonidia. The spheroid on the left is several hours older, and the young spheroids inside have enlarged almost to the size at which they will escape through individual ruptures in the enclosing parent.

CONTROL OF DIFFERENTIATION IN *VOLVOX* 63

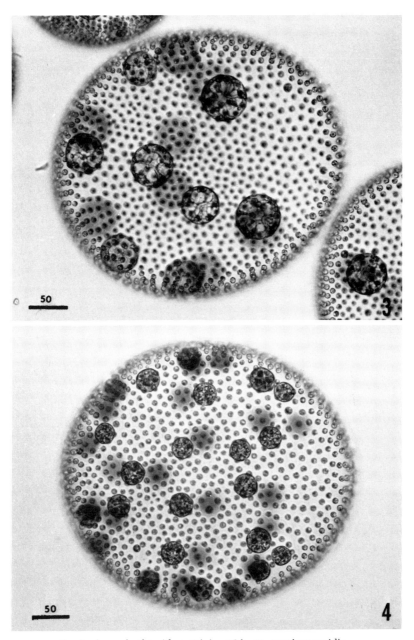

FIG. 3. Asexual spheroid containing 16 large vacuolate gonidia.
FIG. 4. Female spheroid containing approximately 40 small dense eggs.

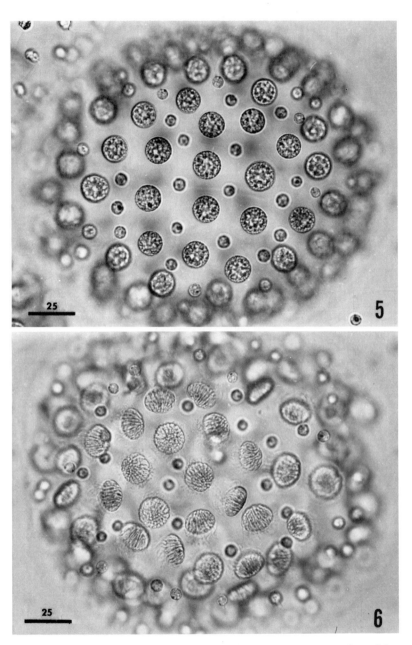

Fig. 5. Male spheroid showing typical 1:1 ratio of small somatic cells and large androgonidia.

Fig. 6. Male spheroid showing sperm packets which have formed by division of the androgonidia. Each packet contains 64 biflagellate sperm.

cultures is added to the medium at the time the young gonidia are enlarging, the pattern of development in the embryo will be altered and a sexual spheroid will be formed. I would like now to describe the sequence of divisions in the embryo which results in each of the three types of spheroids—the asexual, the female, and the male.

The gonidia are formed early in the development of the embryo and thus at the time a young individual is liberated from its enclosing parent, its gonidia which will form the next generation are very evident. As the young spheroid enlarges, two things are especially striking: (1) increase in the size of the spheroid is due mainly to the increase in the size of the cellular envelopes surrounding the protoplasts of the small somatic cells, inasmuch as there is no cell multiplication in the young spheroid after its escape from the parent; (2) although the protoplasts of the somatic cells do not increase in size to any great degree, the asexual reproductive cells, the gonidia, increase from 13 μ up to 90 μ in diameter.

Under the conditions of cultivation used in my laboratory (29°C; 16 hours light/8 hours dark; 600 ft-c intensity), the gonidia usually begin cleavage between 3 and 5 PM on the day after the spheroid in which they are found is released from its parent. Even though gonidia may reach the maturity necessary for cleavage some hours before this, cleavage is routinely deferred until late afternoon under the regime mentioned.

Before considering the details of cleavage, I should like to point out that while we are following the development of an embryo through a series of cell cleavages and movements, the cell in *Volvox* that is being cleaved is haploid and begins cleavage without fertilization.

The pattern of cleavage in the formation of the embryo of *Volvox* has been described in detail by Janet (1923) and Pocock (1933), and more recently by Darden (1966), Kochert (1968), and Starr (1969). These investigators were studying different species and varieties, but in general the basic patterns of the early cleavages are identical in all the species. However, the sequences of growth and cleavage in the developing gonidium and embryo are of two types. In the one (*V. aureus*, *V. globator*, *V. rousseletii*), cleavage begins when the gonidium is very small (Fig. 40), and each cleavage is followed by a period of growth. In the other type (*V. carteri*, *V. africanus*, *V. gigas*), the gonidium undergoes a period of enlargement prior to the onset of the series of cleavages, with no apparent growth between each successive cleavage which may occur as often as every 50–60 minutes. *V. carteri* f. *nagariensis* belongs to the second type.

The gonidia in *V. carteri* f. *nagariensis* are located near the periphery of the spheroid just beneath the single layer of small somatic cells. Every somatic cell and every gonidium is surrounded by a thick envelope of undetermined composition. It is supposed that much of the material is carbohydrate in composition, but as Kochert (1968) has shown, the protoplasts of the somatic cells and the gonidia may be separated by the dissolution of the spheroid using a solution of the proteolytic enzyme Pronase. Gonidia may then be washed with sterile medium using gentle centrifugation and they will continue to undergo their enlargement and subsequent cleavage outside the parental individual.

Early embryogenesis to the 32-celled stage. Development of the new individual from a gonidium (Fig. 11) takes place inside the parental spheroid. Cleavage of a gonidium can be anticipated by a slight flattening of the surface of the gonidium on the side toward the surface of the parental spheroid. This flattening identifies the anterior pole of the early embryo which will result from the series of cleavages that will soon follow. Under the conditions of observation in my laboratory these cleavages occur approximately every 50–60 minutes. The first cleavage is parallel to the longitudinal axis of the gonidium (Fig. 12), and while the second cleavage is also parallel to the longitudinal axis of each cell, each cleavage plane of the second division intersects that of the first at an oblique angle, this resulting in an arrangement of the four cells (Figs. 7 and 13) typical of that seen in the spiral cleavage pattern of the eggs of certain animals. In contrast to the typical cleavage patterns seen in developing embryos of many animals, the third series of cleavages does not

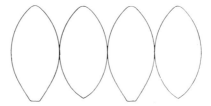

FIG. 7. Diagrammatic representation of the cells from the 4-celled stage in the cleavage of a gonidium of *Volvox* to form a new spheroid. Note that cells 1 and 3 are drawn to indicate that they are the two which touch at the posterior end. After Janet, (1923).

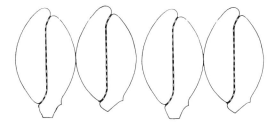

Fig. 8. Diagram of the oblique cleavage of the cells in the four-celled stage, resulting in an anterior tier of 4 cells overlapping a posterior tier of 4 cells. After Janet (1923).

differentiate the animal and vegetal poles. In *Volvox*, this third cleavage is definitely longitudinal but somewhat oblique to the longitudinal axis of the embryo, cutting off in a clockwise fashion an anterior tier of cells (Figs. 8 and 14). The 8-celled embryo consists, therefore, of two tiers of cells, the anterior tier shifted clockwise to alternate with the lower tier, which it overlaps considerably (Fig. 15). The 4 cells at the posterior end of the embryo remain in close contact, and while the 4 anterior cells may move anteriorly to form the typical hollow sphere, a small pore, the phialopore, remains evident between the anterior cells. The orientation of each cell in the 8-celled stage has become such that the nuclear end of the cell is pointed inward. All further divisions will continue to be parallel to the longitudinal axis of each cell (and thus parallel to the radial axes of the embryo), but the positioning of the cleavage planes in relation to the anterior-posterior axis of embryo occurs in a repeating pattern which has been painstakingly described by Janet (1923).

Even though the 8-celled stage consists of an anterior tier of 4 cells overlapping a posterior tier of 4 cells, the establishment of the basic animal and vegetal poles of the embryo does not occur until the fourth cleavage, at which time the cells of the 8-celled stage divide to form 16 cells of equal size, arranged in four alternating tiers of 4 cells each (Fig. 16). In the division of the 8-celled stage, the anterior 4 cells divide obliquely to form two tiers of 4 cells each; the posterior 4 cells form two tiers of 4 cells in a similar manner (Fig. 9). However, the overlapping position of the original anterior and posterior tiers in the 8-celled embryo results in the production of a 16-celled embryo in which tiers 1 and 3 originate from the anterior tier of the 8-celled stage while tiers 2 and 4 are derived from the posterior tier (Fig. 10). The anterior part of the young embryo will develop from

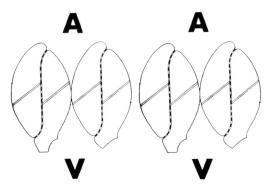

Fig. 9. Diagram showing the position of the fourth cleavage in the embryo, resulting in the delimitation of the animal (A) and vegetal (V) poles. After Janet (1923).

tiers 1 and 2 while the posterior region of the young embryo will be derived from tiers 3 and 4. The animal and vegetal poles have now been established. The divisions to form the 32-celled embryo are in such relationship as to maintain the spherical form of the developing embryo and the basic positions of the cells relative to the animal and vegetal poles (Janet, 1923). In *V. carteri* f. *nagariensis* the development of all types of embryos—asexual, female, and male—is the same through the formation of the 32-celled stage (Fig. 17).

Development of the asexual embryo. In both the male and the female strains the development of the asexual embryo under optimum growth conditions is identical. In the development of the 32-celled embryo, each successive cleavage has resulted in cells of smaller, but equal, size. It is at the division of the 32-celled stage of a developing asexual embryo that the first morphological differentiation is apparent. At this division the anterior 16 cells (the animal pole) of the embryo undergo unequal cleavage, forming a small somatic initial and a large gonidial initial (Fig. 18). Under growth conditions less than optimum, there may be a reduction in the number of cells which will show this differentiating division and the resulting embryo may have fewer than 16 gonidia, usually 10 to 15.

Further divisions occur in the embryo in apparent synchrony. The large gonidial initials may cut off small cells (Fig. 19), but it is doubtful that this continues for many divisions. The somatic initials in the anterior and posterior parts of the young embryo continue their synchronous divisions; and as the cells become smaller and smaller with each division, the gonidia appear relatively larger and more distant from each other, retaining always their original positions in the anterior half of the young embryo. Cell divisions cease in the embryo usually after the 11th or 12th division, and thus the final cell number in the spheroid under optimum conditions is usually between 2000 and 4000 cells. The cessation of cell divisions is most probably a function of the size of the dividing cells inasmuch as smaller gonidia from cultures grown under less than optimum conditions can be expected to form embryos with fewer cells.

At the termination of cell divisions, the spheroidal embryo has its many somatic cells with their nuclear ends (and subsequently flagellar ends) pointing inward; the 16 or fewer large gonidia project from the outside surface. Soon, however, in the region of the phialopore two intersecting cracks appear in the embryo, and the process of inversion begins. The four lips bend back over the surface of the embryo, and within 45 minutes the embryo turns itself insideout (Figs. 20–22). The mechanism for this process is unknown, but it is certain that it does not involve cell multiplication. The stubs of the pair of flagella on each somatic cell become evident as inversion proceeds, but in the young inverted embryo only one member of each pair of flagella grows to its maximum length at first. Prior to release of the young embryo, the second flagellum reaches its maximum length.

The anterior pole of the developing embryo with its prominent gonidia becomes the posterior pole of the young spheroid after in-

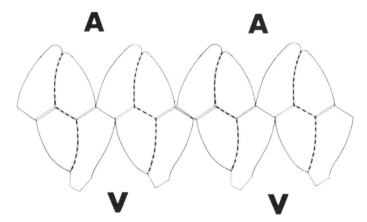

FIG. 10. Diagram of the young embryo after the fourth cleavage and the enlargement of the 16 cells. The 8 cells at the animal pole (A) which have been delimited by the fourth cleavage will give rise in the sixth cleavage to the asexual reproductive cells of the developing embryo. After Janet (1923).

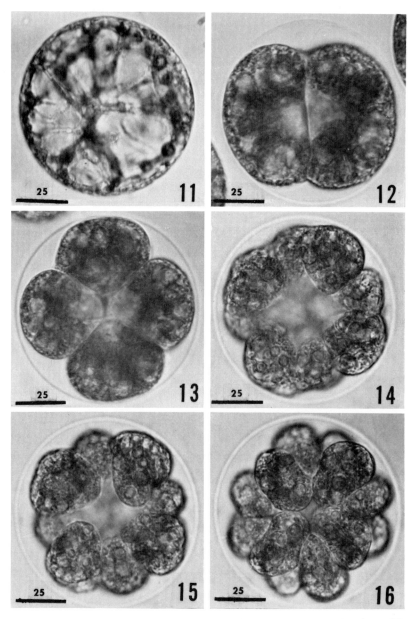

FIGS. 11–22. Stages in the cleavage and development of an asexual embryo. The series is not a sequence from a single gonidium. The gonidia were removed from the parental spheroid and allowed to develop outside in order to facilitate photography.

version. The gonidia remain in the same relative positions where they were formed during early embryogeny.

Development to the female embryo. Under the influence of an inducer produced by sexual males, the gonidia of the female strain will form egg-bearing spheroids (Fig. 4). In the development of the egg-bearing embryo by the female strain the cell divisions to form the 32-celled stage are identical with those described earlier for the asexual embryo, but unlike those of the asexual embryo, the cell divisions in the 32-celled stage are equal and form a 64-celled embryo in which all cells are the same size. Differentiating cell divisions occur when the 64-celled stage divides. Usually not only the anterior 32 cells divide unequally, but some of the more anterior cells of the posterior half also do so, this resulting in embryos with 35–45 egg initials as a rule. Further development of the egg-bearing embryo follows the pattern of cleavage and inversion which occurs in the asexual embryo.

Development of the male embryo. The gonidia of the male strain respond to inducer from sexual males by forming male embryos. Cleavages in the developing male embryo follow the typical pattern of the asexual embryo to the 32-celled stage. In the male embryo further divisions to form 64, 128, and 256 cells are equal, and under optimum conditions a final division to form 512 cells can be expected. This final division is unequal in all the cells of the embryo and results in a male embryo with 1:1 ratio of large androgonidia (sperm-producing cells) and small somatic cells (Fig. 5). Under less than optimum conditions the final division may occur at an earlier stage in the embryo, but whether the final division is at the 32-, 64-, or 128-celled stage, the cleavage of every cell is unequal and the typical 1:1 ratio of large and small cells is evident. After inversion, the

---

FIG. 11. Enlarging gonidium. Gonidium will become dense prior to onset of a cleavage.

FIG. 12. Two-celled embryo. Note enclosing cell wall (?) of gonidium.

FIG. 13. Four-celled embryo showing typical arrangement of cells following spiral cleavage, 2 cells touch while the other 2 fit into angles at base.

FIG. 14. Cleavage of the 4-celled embryo as seen from the anterior end. Note that each cell is cutting off obliquely in a clockwise direction a slightly anterior cell.

FIG. 15. Later stage in same embryo seen in Fig. 14, showing 8-celled embryo with anterior tier of 4 cells alternating with the posterior tier. Embryo seen from anterior end. If viewed from side, the two tiers would be seen to overlap; see Fig. 8.

FIG. 16. Sixteen-celled embryo resulting from cleavage of 8-celled embryo seen in Fig. 15. Four tiers of 4 cells alternate; compare with Figs. 9 and 10. The small pore between the 4 cells at the anterior end is the phialopore.

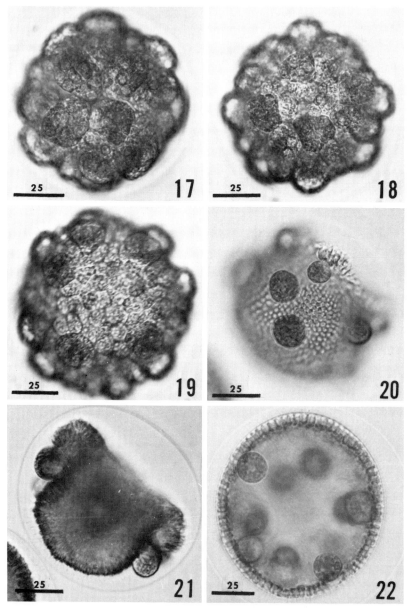

FIG. 17. Thirty-two-celled embryo. Note 4 cells surrounding phialopore. All cells in embryo are equal in size.

androgonidia enlarge and undergo a series of cleavages to form packets of 64 or 128 biflagellate sperm (Fig. 6).

Release of the Young Spheroids

The young spheroids of all types begin enlargement while still in the parental individual; but before reaching a size which would result in compression of the spheroids by the enclosing parental cells, they are released singly through ruptures in the parental wall. The nature of this release mechanism is not understood. One often observes that in aerated cultures spheroids which would have remained for some time will be released within a few minutes after aeration has ceased; this happens, of course, only in cultures that have reached the proper state of maturity.

Sexual Fusion and Zygote Germination

The eggs in the female spheroids are receptive for some hours after the females have been released. If placed in fresh *Volvox* medium, the eggs will begin to enlarge within 24 hours and will revert to asexual reproductive cells, undergoing cleavage to form new embryos; on the other hand, work in progress indicates that the presence of inducer in the medium will inhibit this return to the asexual function.

The male spheroids contain fully developed packets of 64 or 128 sperm at the time of their release from the parental individual. The released male is very ephemeral and within a few hours dissociates, releasing the intact sperm packets. Sperm packets swim in the medium and upon contact with a female spheroid, dissolve a hole in the female. The sperm packets dissociate and the individual naked, elongate, biflagellate sperm burrow through the female to the many eggs. The details of the sexual process have been reported by Starr (1969), but at that time fertilization had not been observed. More recently, using a female mutant with many more eggs than normal and with

---

FIG. 18. Sixty-four celled embryo showing gonidia initials formed by unequal cleavage of the anterior cells in the embryo at the division of the 32-celled stage.

FIG. 19. Older embryo with 256 cells. Gonidia become more evident as successive cleavages result in smaller somatic initials.

FIG. 20. Embryo at beginning of inversion. Lips surrounding the philapore area are bending back over the surface of the embryo. All cell divisions are now complete.

FIG. 21. Embryo halfway through inversion.

FIG. 22. Inverted embryo with gonidia now on inside of spheroid and flagellar ends of somatic cells pointing outward.

fewer somatic cells to obscure observation, fertilization has been observed. Details will appear in a later account.

Within 3–4 days after fertilization the zygotes, still enclosed in the female spheroid, will have become orange in color, secreted a thickened crenulate wall, and entered a period of dormancy. The zygotes can be induced to germinate after approximately 3 weeks of dormancy by transferring them to the fresh *Volvox* medium. Each zygote produces a single biflagellate zoospore which contains a single viable haploid nucleus resulting from the meiotic division of the zygote nucleus on germination. The zoospore will swim for a short time and then undergo cleavage to form a small germling individual which usually contains 8 gonidia. Details for the germination process have been given in Starr (1969).[1]

The Inducer

It has been shown clearly in kinetic experiments that the inducer is not present in any reasonable strength in a population of sexual males until after the release of the sperm packets and their eventual disintegration. Whether the inducer is a secondary metabolite produced by the sperm or the somatic cells of the male, or a disintegration product of one or both of them is not yet known.

Assay. The assay of the inducer is carried out using the female strain (HK 10) inasmuch as it has an absolute requirement of the active principle for the formation of egg-bearing spheroids. Details of the assay method have been published by Starr (1969). Serial dilutions of the fluid to be tested are made by successive transfers of 1 ml in a series of 9 ml media blanks in 18 × 150 mm tubes. The dilutions are inoculated with five spheroids of the female strain just prior to the time the young asexual spheroids are to be released from the parental spheroid. As a result, each dilution tube will contain on the average 75 young asexual spheroids (each with 14–16 gonidia) and thus can be expected to produce 1000+ offspring in each tube which are scored as asexual or female to determine the percentage of induction in a given assay tube. In those dilution tubes where the inducer is not limiting, 100% (or nearly so) induction is obtained. Only one tube of a series usually contains a mixture of asexual and egg-

---

[1] A 25-minute, 16 mm, color film entitled "*Volvox*: Structure, Reproduction and Differentiation in *V. carteri* (HK 9 & 10)" has been made in cooperation with Dr. C. M. Flaten of the Audio-Visual Center, Indiana University. This film (no. FSC-1257) is available for sale or rental from the AV Center.

bearing spheroids, indicating this to be the dilution at which the inducer is limiting; the tube with the next highest dilution will show no induction at all.

Potency. Batches of male fluid with high potency can be obtained through the induction of large synchronous population of a newly isolated recombinant of the two wild-type strains. When the original wild-type male (HK 9) was used, fluids with 100% activity at a 10^5 dilution and 50% at 10^6 were the most potent obtained. With the new recombinant, fluids produced in a similar manner give 100% induction at 10^7 and 50% at 10^8, an increase of 100-fold in potency. Furthermore, by concentrating the parental spheroids just prior to the release of the males into the culture medium, it is now possible to produce fluids giving 50% induction of 10^9 dilution.

The potency of the fluid produced by the sexual males can be more easily appreciated when one estimates the number of gonidia of the female strain which can be induced as a function of the number of sperm produced in the male culture. Estimates were made by inoculating 10 washed parental spheroids containing sexual males into tubes containing 10 ml of *Volvox* medium. After allowing the males to be released and their sperm packets to disintegrate, the tube was assayed for inducing activity. When using the wild-type male, activity was 100% at the 10^3 dilution, with some activity at 10^4; with the new recombinant male, activity was 100% at 10^5 dilution and 50% at 10^6. If in the latter case we consider that the activity in 1 ml of the original tube was the result of the action of a single parental spheroid with its males, we may make the following calculation, using maximum values: 1 parent $\times$ 16 males $\times$ 256 sperm packets $\times$ 64 sperm = 262,144 sperm. One milliliter of the inducing fluid diluted 1,000,000 times resulted in the production of 500 females among 1000 assay organisms in a 9-ml tube, thus 55,555,555 females would be formed if the 1,000,000 ml were used. Thus, 55,555,555 gonidia could be induced by the action of a single parental spheroid the offspring of which formed 262,144 sperm, i.e., for each sperm produced, 212 gonidia can be induced.

Purification. Concentration of the active principle from the male fluid has been effected by adsorption on a cation exchange column of carboxymethyl cellulose (Bio-Rad Laboratories), elution with 0.1 M NaCl in 0.05 M citrate phosphate buffer at pH 5, and concentration by flash evaporation. Further concentration and purification were achieved by using Sephadex G-75, which allows for the fractionation of globular proteins of 3,000–70,000 molecular weight. The active principle was eluted in a single peak in fractions between those in which ovalbumin (MW 45,000) and α-chymotrypsin (MW 25,000) were eluted. Lyophilization of the active fractions from the Sephadex yields a dry fluffy powder with very high activity. The highest activity obtained in assays of this lyophilized extract was 100% induction at a concentration of 10^{-10} gm/liter, i.e., 0.1 ng/liter, but routine extractions usually give 100% activity at 10^{-9} gm/liter with 50% activity at the 10^{-10} gm/liter level. Assuming a molecular weight of 30,000, the activity of the routine extractions would be 50% at a concentration of 3×10^{-15} M.

The activity of the raw male fluid and of the concentrated inducer is much reduced by the proteolytic enzyme Pronase, but the activity is not affected by trypsin, chymotrypsin, ribonuclease, deoxyribonuclease, or phosphodiesterase.

The absorption spectrum of the lyophilized material with activity at the 0.1 ng/liter level showed both a protein and a nucleic acid peak; after digestion by purified phosphodiesterase and dialysis with several changes of buffer, the nucleic acid peak was removed but the inducing activity of the solution remained.

Specificity. All attempts to effect cross-induction between the HK strains of *V. carteri* f. *nagariensis* and other isolates of the same species (but a different morphological form, *V. carteri* f. *weismannia*) have proved unsuccessful. Kochert during his predoctoral study in my laboratory had a number of strains of the f. *weismannia* from various locations in the United States and found that the female strain from Nebraska, which he studied most closely (Kochert, 1968), could be induced by males from other locations (Indiana, New Jersey, and elsewhere); since that time I have isolated other clones of f. *weismannia* from Texas, North Carolina, and Indiana which show cross-induction with the Nebraska strains. On the other hand, a population of *V. carteri* f. *weismannia* from Australia will not cross-induce with the Nebraska strains. Clones from India gave the most surprising result by cross-inducing with both Australian and Nebraskan strains (Fig. 23).

Effects of carbon dioxide and illumination on the response to the inducer. The difficulty of achieving induction at the 100% level in the female Nebraskan strain (Kochert, 1968), as well as other female strains of *V. carteri* f. *weismannia* isolated from the United States was not understood at first; in fact, for most of the time Kochert was work-

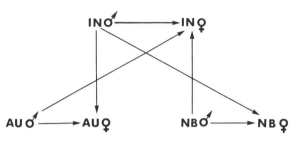

Fig. 23. Pattern of cross-induction of female strains by inducing fluid from males. The HK clones are *Volvox carteri* f. *nagariensis* from Japan while the other strains (Australia, India, and Nebraska) belong to *V. carteri* f. *weismannia*.

ing in my laboratory the difficulty was not even recognized inasmuch as the HK strains of *V. carteri* f. *nagariensis* had not at that time been isolated and shown to give 100% induction at dilutions in which the inducer was not limiting. In April, 1970, from the HK 10 female strain of *V. carteri* f. *nagariensis* I isolated a mutant strain in which the response to the male inducer was markedly different from the normal. In dilutions of the inducer containing as much as 10,000 times the quantity of inducer necessary to effect female production in the normak HK 10 strain, the mutant strain (Starr 70-3) rarely gave more than 2% females. Many of the asexual spheroids in such tubes, however, showed a slight increase in the number of gonidia, having 18 to 22 rather than the maximum of 16 found in noninduced populations of both the normal and the mutant strain. In dilution series this type of response was evident in all dilutions in which corresponding tubes inoculated with the normal strain produced females. Thus, it was evident that the inducer was getting into the cells but that its action was being at least partially blocked or neutralized in some manner. Further experimentation showed that by increasing the level of CO_2 in the atmosphere surrounding the assay tubes, it was possible to increase the production of females in the mutant strain to as much as 25% of the 1000+ individuals in a single assay tube. At the same time many of the asexual spheroids in the induction tubes had gonidia in numbers equal to the eggs in a typical egg-bearing spheroid (Fig. 26), while a few spheroids contained both eggs and gonidia (Fig.

27), a condition never observed in the normal strain. Duplicate dilution series of the inducer inoculated with the normal HK 10 female showed no increase in the sensitivity of the normal female. Increasing the level of CO_2 above 0.5% in an effort to increase the number of females in the mutant strain was not successful because the buffering system of the dilute medium was not adequate to prevent the lowering of the pH to such an extent that growth was inhibited. The pH of the medium is lowered to pH 6 by the 0.5% CO_2, but media at pH 6 or pH 6.5 containing inducer did not have the effect of increasing the number of females in the mutant strain. Whether the increased CO_2 has its effect on a defect in some biochemical process directly related to the formation of eggs in the mutant female or whether the increase in CO_2 is partially counteracting some defect in the photosynthetic system is under continuing investigation.

The female Nebraska strain and a second female strain from North Carolina were tested to see whether an increased level of CO_2 would have any effect on the induction process induced. When induced under the higher intensity of illumination (600 ft-c) and in an atmosphere of 0.5% CO_2, the response of both females was similar to that regularly obtained for the HK strains of f. *nagariensis*, i.e., a 100% formation of females in those dilutions were the inducer was not limiting. Induction under lower intensity of illumination of 350 ft-c (Kochert, 1968) with or without the increased CO_2 gave erratic results. Whether the role of the CO_2 in this instance is the same as in the mutant HK 10 female is unknown, but its possible role in increasing the photosynthetic rate and thus the formation of sexual spheroids is a very tempting hypothesis in both cases.

Kochert (1968), in his study of *V. carteri* f. *weismannia*, was concerned only with the induction of females by the inducer from the male, pointing out that the male strain formed the typical dwarf male individuals spontaneously, usually 2 or 3 generations after transfers were made into fresh medium; the culture then would produce only asexual individuals until a transfer to fresh medium was made again. Thus, the inducer was being produced at a time when the male culture was very sparse, making it impossible to get inducing fluids of high potency. As indicated earlier, my subsequent isolation of the HK strains of *V. carteri* f. *nagariensis* in which the male strain was easily inducible appeared to provide us with unusual material in which male fluids with high induction potency could be obtained and at the same time provide a second pattern of sexual differentiation for analysis. Under increased levels of CO_2 and illumination, we find that the Nebraska male can now be induced at will, and male fluids of high potency obtained. In contrast to the fluids of the sexual Nebraskan male reported by Kochert as having activity at a dilution of 10^3, but not at 10^4, induced males of the same strain have yielded now under the CO_2 treatment fluids with activity at a dilution of 10^6. Assays of this inducer using the Nebraskan female strain, the Carolina female strain, and the Nebraskan male gave comparable results. The earlier observation that the Nebraskan strain formed males shortly after transfer to fresh medium can be explained as the response of the inoculum to inducer carried over with it from the old population at the time when CO_2 was not completely limiting. The pattern of cross-induction between the males of the different populations of f. *weismannia* can now be examined.

Attempts to separate mixtures of the active principles from f. *weismannia* and f. *nagariensis* have not been successful. With electrophoresis, using cellulose acetate strips, excellent migration of the inducer could be obtained, but no separation of the different inducers was evident when the strips were cut into very narrow bands, eluted in Volvox medium, and assayed in a dilution series using females of the two types as assay inocula. Fractionation of the inducing fluids from the several male strains could be effected with Sephadex G-75, but the active principle in each strain of the f. *weismannia* came off in the same fractions as the inducer from the f. *nagariensis*.

Initiation of the Sexual Response

Populations of the male strain in *V. carteri* f. *nagariensis* (Starr, 1969) were reported to go sexual "spontaneously" with the production of a few males followed by the successive induction of a large number in the next generations, but the mechanism by which this *de novo* production of males was brought about has not been explained. Some insight into the mechanism of control might be inferred from study of the female strain.

Spontaneous females. There were occasional observations in my laboratory that the female strain of *V. carteri* f. *nagariensis*, in spite of an "absolute" requirement of inducer from the male in order to form egg-bearing spheroids, did in fact occasionally produce egg-bearing spheroids "spontaneously" in very low frequency (less than 1 in 10,000). Inasmuch as egg-bearing spheroids of *V. carteri* f. *nagariensis* do not produce any inducer, there was no effect on the remainder of the population in which they might appear. It was assumed earlier that such spontaneous females were merely biochemical mistakes of little interest, but more recently in trying to find an explanation for the appearance of the spontaneous males in the male strain, a more detailed investigation of the spontaneous female was begun, with the following question in mind: Are spontaneous females the result of ephemeral biochemical mistakes on the transcriptional or translational level or are they the result of permanent changes at the gene level?

The first evidence that the spontaneous females were different from females resulting from induction by the male inducer was the difficulty encountered in establishing clonal cultures from these spontaneous females. Cultures can be established routinely from induced females by transferring to fresh Volvox medium. Within a few days after transfer, all the eggs begin to enlarge and subsequently undergo a series of cleavages followed by inversion to form small asexual spheroids. Asexual spheroids will continue to be formed in successive generations, and only under the influence of the inducer from the male will egg-bearing spheroids again be produced. When spontaneous females were isolated and placed in fresh Volvox medium, the eggs remained dark green, and after a few days, they disintegrated. Various media on hand were tried, and Volvox medium containing sodium acetate stimulated the growth of the eggs. The sodium acetate was reduced to the level of 0.025% as it was somewhat toxic at higher levels even to the normal asexual spheroids. A number of clones have been established from the spontaneous females. Some of the clones continue to have an apparent requirement for low levels of sodium acetate, while others show only a slight stimulation, if any. Most surprisingly certain of the clones have continued to produce only egg-bearing spheroids in all subsequent generations; in such clones the eggs enlarge, undergo cleavage, differentiate eggs, invert and are released as typical females. This cycle is repeated in successive generations, but without the presence of any inducer from the male. Another spontaneous female clone (Starr, 70-7) which has been studied more closely will under less than optimum conditions produce some asexual spheroids, but when placed again under optimum growth conditions females are produced without inducer being present. Assay of the fluid from such clones of spontaneous females does not reveal the presence of any inducer.

In the short time since we were first able to establish cultures from spontaneous females, we have been able to make only a first attempt to see the pattern of inheritance through the sexual cycle. The spontaneous female clone (Starr 70-7) was crossed with a normal male (Starr 69-1b), and the progeny from zygotes were analyzed. In *Volvox*, meiosis occurs at some time before or during germination and only one haploid product survives. The analysis of the over 300 haploid products gave a near 1:1 ratio of males to females, but the most significant result was the observation that all female clones were *spontaneous* females, while the males were typical males.

In answer to my original question as to the nature of the spontaneous females, it would seem, indeed, that they are permanent alterations at the gene level and that, in the one case analyzed, the change is linked to the sex locus. Whether a similar alteration does actually operate in the formation of spontaneous males can only be inferred, but the possibility is most intriguing.

Site of action of the inducer. The site of action of the inducing principle and its biochemical role in changing the pattern of development from the asexual to the sexual phase are problems for the future. Whether the inducer acts inside the cell or only on the cell surface awaits experiments with purified labeled inducer. There is, of course, the usual temptation to explain the action of the inducer in terms of the Jacob–Monod operon model in which the inducer would bind the repressor and thus allow the operator gene of the sex operon to function; this becomes even more tempting when we consider the spontaneous female mutant which is permanently turned-on due to a mutant locus on the same chromosome as the sex locus. On the other hand, the so-called inducer with its estimated 30,000 MW may not be an inducer after all. Could this active principle, harvested from male culture fluid in which the sperm have disintegrated, be a repressor of the asexual system, and through its action on the asexual system eliminate the repression of that system on the sexual phase?

Mutants Affecting the Developmental Pattern

Examination of the stocks of the wild-type male of *V. carteri* f. *nagariensis* in September, 1969, some 30 months after its isolation into a clonal culture, showed that the strain was no longer capable of producing the typical sexual males with a 1:1 ratio of androgonidia and somatic cells. From the stocks, clones of two different mutant males were isolated; both mutants had reduced numbers of androgonidia and concomitantly an increased number of somatic cells. The one mutant (Starr R-1) shows a change in the time of cellular differentiation; in the development of the sexual male embryo, unequal divisions to

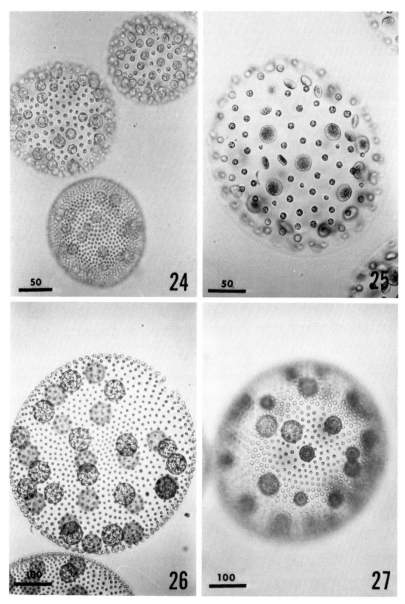

Fig. 24. Male spheroids of the R1 mutant of *Volvox carteri* f. *nagariensis*. Compare with normal male seen in Fig. 5.

Fig. 25. Male spheroid of the R-2 mutant of *Volvox carteri* f. *nagariensis*. Note very

form the androgonidia occur earlier than the last division and thus result in fewer androgonidia. At times, the R-1 male resembles very much the egg-bearing female in its relative numbers of cell types (Fig. 24).

The R-2 mutant (Fig. 25) exhibits a decrease in the number of cells participating at the last division of the male embryo, which is typically an unequal division to form the large reproductive cells. Among the larger numbers of somatic cells in the male spheroid of R-2 strain, one can easily distinguish those somatic cells which result from the last unequal division to form an androgonidium and a somatic cell; such cells appear much smaller and are located next to the androgonidium which was formed at the same time by the unequal division.

Crosses involving the R-1 male mutant with the normal HK 10 female strain show that the mutant trait is inherited as a single factor, with approximately half of the male progeny producing mutant males, while the other male progeny produce normal males with the 1:1 ratio of androgonidia and somatic cells. The effect of the mutant trait in the females, half of which were probably carrying the trait, was not immediately apparent and at the time was not investigated further.

The R-2 mutant male was also crossed with the HK 10 female and the progeny from the germinating zygotes analyzed. Although there was good germination, approximately 50% of the small germling spheroids failed to survive. The germlings which did not survive enlarged in the typical manner but their reproductive cells did not enlarge and thus did not produce any offspring. In the 50% which did survive, the mutant trait segregated independently of the sex locus, resulting in offspring of four different types: normal males, mutant males, normal females, and mutant females. The females carrying the R-2 mutant trait are characterized by a reduced number of eggs; the number of the eggs in the posterior part of the spheroid (formerly the anterior part of the embryo) was in normal frequency, but the anterior part of the spheroid was at times completely devoid of eggs. Thus, it seems that the R-2 mutant trait reduces the number of cells in both

---

small somatic cell by each large sperm packet. Occasional pairs of small packets result from equal division at last cleavage with potential to form sperm.

FIG. 26. Asexual induced spheroid of the "sterile" female mutant (Starr 70-3) of *V. carteri* f. *nagariensis*, showing multiple gonidia instead of eggs.

FIG. 27. Induced spheroid of the sterile female mutant (Starr 70-3) with both eggs and gonidia.

male and female strains which are involved in the differentiation of the sexual cells. The asexual spheroids of both male and female strains bearing the mutant trait do not show any reduction in numbers of gonidia.

The genetic basis for the lethality of half of the offspring of the original cross becomes even more puzzling because intercrosses of male and female offspring (mutant and normal in all combinations) give 100% survival, but back-crossing of female offspring with the original R-2 mutant male gives varying degrees of germling lethality, sometimes approaching the 100% level. Further investigations will be made of this phenomenon.

While examining a culture of the normal HK 10 female, a spheroid was observed with small gonidia, lighter in color than the normal and numbering above 20. This spheroid was isolated, and the condition proved to be stable. The mature spheroid may have from 16 to 40 or more gonidia, some single, some in pairs or triplets, occupying the same relative positions in the spheroid as the 16 gonidia do in the normal spheroid (Fig. 28). The origin of these multiple gonidia during the early developmental stages has not yet been studied. Analysis of a few germlings from a cross between the mutant (Starr 70-5) and a normal male shows that the trait is passed through the sexual cycle. A male recombinant carrying the mutant trait has been isolated. The egg-bearing spheroids are not obviously different in cell arrangement and numbers in the female strain bearing the mutant trait, but in the male mutant strain, the male spheroids do not show the regularity of pattern observed in the normal. Some males will have more somatic cells than expected, while many males have almost no somatic cells, being composed almost entirely of androgonidia.

The mutants described above, as well as others which are being investigated, indicate that embryogenesis in *V. carteri* is under a series of genetic controls. Our knowledge of the nature of these controls which determine the time of differentiation, the site of the differentiating divisions, and the ultimate development of the differentiated cells will hopefully increase with further study of these and other mutants in the future.

The Determination of Somatic Cells

In the preceding sections we have been concerned mainly with the events involved in the differentiation of the reproductive (germ) cells.

It should be emphasized that the somatic cells in *Volvox* have also undergone differentiation and are no longer capable of resuming division after the inversion of the young embryo. In multicellular animals, with few exceptions, such as the Rotifera, somatic cells may continue to grow and divide during the life of the individual. In *Volvox*, the only cells that retain the ability to resume division are the reproductive cells—gonidia, androgonidia, eggs—and even in the eggs, this is suppressed during the period when the egg is receptive to the sperm. I have mentioned earlier the extreme suppression of growth and division processes in eggs of spontaneously formed females, a suppression that could be overcome with the addition of sodium acetate to the medium; however, sodium acetate will not overcome the suppression in the somatic cells.

In the algal family Volvocaceae, the colonial genera *Pandorina*, *Eudorina*, *Pleodorina*, and *Volvox* form an interesting series of spheroids showing an increase in cell number and in the differentiation between somatic cells and reproductive cells. In *Pleodorina* (*Eudorina*) *illinoisensis* Kofoid (Fig. 29), the simplest member of the series showing differentiation of somatic and reproductive cells, the four anterior cells of the 32-celled spheroid are small and somatic, the remainder being reproductive (Kofoid, 1898; Goldstein, 1964); while in *Pleodorina californica* Shaw (Fig. 30) almost one-half of the 128-celled spheroid may be composed of somatic cells. These somatic cells arise, as has been seen in *Volvox*, from the posterior cells in the young embryo, i.e., the centermost cells of the dividing embryo (Goldstein, 1964; Gerisch, 1959). The control of the production of these somatic cells is unknown, but certainly the change is not a permanent one in the genome inasmuch as Gerisch (1959) was able to get occasional regeneration of a somatic cell in *Pleodorina californica* to form a new little spheroid from which normal populations arose.

In April, 1970, while searching for spontaneous females in the asexual cultures of the female strain of *V. carteri* f. *nagariensis*, I came across a parental spheroid which had released most of its offspring but which had a patch of some 50 to 60 very small spheroids closely packed in a single layer on one side of the remaining somatic cell mass. The small spheroids were isolated into tubes of media where they enlarged, but in a manner quite different from the normal spheroids of the female strain. The somatic cells of the small spheroids began to enlarge with little production of the gelatinous cellular matrix except in the most anterior region of the spheroid

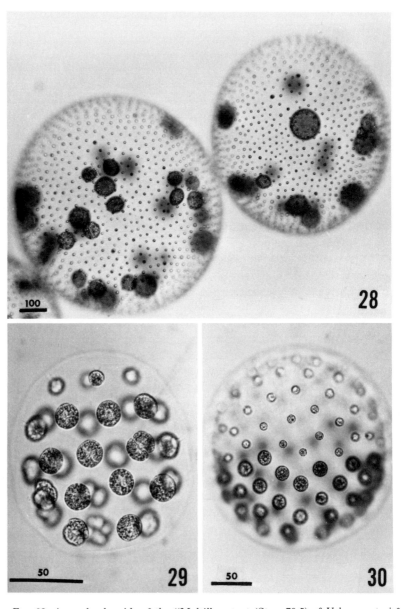

Fig. 28. Asexual spheroids of the "Multi" mutant (Starr 70-5) of *Volvox carteri* f. *nagariensis*. Note formation of multiple offspring in groups in spheroid on the left. Compare with normal asexual spheroid in Fig. 2.

Fig. 29. *Pleodorina (Eudorina) illinoisensis*. Side view of spheroid showing anterior tier of small somatic cells. Some cells in posterior are beginning cleavage.

Fig. 30. *Pleodorina californica*. Side view of spheroid showing small somatic cells in anterior half.

(Fig. 33). The gonidia enlarged as usual and subsequently began to cleave to form new individuals; the enlarging somatic cells also began to cleave, and, depending on the size they had reached at the time the cleavage was initiated, they formed small spheroids varying in size from 16 cells upward to several hundred. The new small spheroids differentiated in the usual manner producing large gonidia (sometimes only one or two); and, upon release, each small spheroid went through a period of growth and enlargement, followed by cleavage of both gonidia and enlarged somatic cells. The somatic mutant is stable and continues to produce spheroids in successive generations in which somatic cells have remained reproductive.

When the somatic mutant was grown in medium containing inducer, the large gonidia as well as the somatic cells produced egg-bearing spheroids (Fig. 31 and 34). Even the somatic cells of the egg-bearing spheroids appear to transform slowly into cells with dense cytoplasm such as is observed in eggs, but such transformed cells have not yet been shown to be capable of fertilization. When crossed with the normal male, zygotes are formed and can be germinated in the usual manner. Of the 65 progeny analyzed, 31 were capable of somatic regeneration and 34 were strains with sterile somatic cells, an indication that the genetic basis for the somatic trait is probably chromosomal. The sex of the progeny was not determined, but a recombinant bearing the male locus and the trait for somatic regeneration was identified and isolated.

The male strain bearing the somatic mutant trait can be induced to form male spheroids (Figs. 35 and 36). Both the gonidia and the somatic cells were induced to form typical male spheroids with the characteristic 1:1 ratio of androgonidia and somatic cells, but in this case the somatic cells of the male are not sterile; they form small packets of sperm at the same time the androgonidia form large packets (Fig. 32). The observation that such somatic cells in the male regularly form sperm packets would seem to indicate that the biochemical processes necessary for the production of sperm have been initiated in the cells of the male embryo prior to the time of the unequal divisions of the cells at the last division in the embryo. However, the occasional observation that a sperm packet may be formed directly from an enlarging somatic cell of an asexual spheroid placed in inducer would caution against accepting this conclusion without more convincing biochemical evidence. The isolation of a mutant the somatic cells of which are no longer sterile, does not, of course, ex-

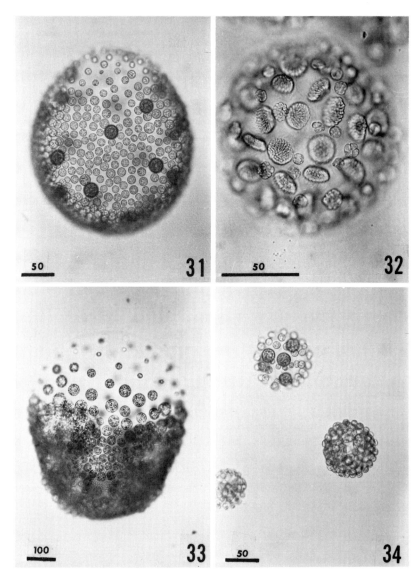

Figs. 31–34. Spheroids of *Volvox carteri* f. *nagariensis* bearing the genetic trait which results in the formation of fertile somatic cells.

Fig. 31. Female spheroid showing eggs. Note that somatic cells are enlarging, but little gelatinous matrix has been secreted around each one.

Fig. 32. Male spheroid showing the formation of large sperm packets by the androgonidia and the formation of small sperm packets by the somatic cells.

plain why the normal somatic cells become incapable of further division or why those in the mutant can divide, but it does provide us with material for future exploration of the events in both conditions.

VOLVOX AUREUS EHRENBERG (FIGS. 37–40)

Induction of Males

The first induction system in *Volvox* was discovered by William Darden while a student at Indiana University, and in his doctoral thesis (Darden, 1965), and his subsequent paper (Darden, 1966), he outlined a system in which the medium from a sexual male culture would induce males (Fig. 38) to form in cultures of *V. aureus* which otherwise would produce only asexual spheroids (Fig. 37). *Volvox aureus* does not form special females with eggs; the asexual reproductive cells act as facultative eggs. Thus, one is limited to a male self-inductive system in this species.

Darden reported that even in tubes of medium containing a large excess of the inducer (as indicated by activity of subsequent dilutions in the series) "male colonies were rarely found to exceed 50% of the total colonies present in a given culture." In 1965, there was little basis for questioning the conclusion that a maximum induction figure in *V. aureus* might lie around the 50% level. However, with the subsequent isolation and study of *V. carteri* f. *nagariensis* (Starr, 1969), *V. rousseletii* (McCracken and Starr, 1970), and *V. gigas* (Vande Berg and Starr, 1971), in which dilution series of the active male fluid give 100% induction in those tubes of the series where the inducer is not limiting, the concept of a 50% maximum for *V. aureus* becomes less tenable. Experiments in my laboratory have shown that the 50% maximum is indeed a reflection of the method by which the assay is conducted. Darden (1966) had reported that the most susceptible stage of the young embryo for induction was at a period just before the end of cell division and the beginning of inversion; in his experiments, this was the 48-hour developmental period. Darden's use of newly released spheroids as inoculum for his assay meant that the most susceptible stage of the embryo would not be reached for 48 hours. In dilution series set up

FIG. 33. Asexual spheroid showing sterile somatic cells in the most anterior region. Sterility is accompanied by an increase in the amount of gelatinous matrix secreted by each somatic cell.

FIG. 34. Small spheroids formed by cleavage of somatic cells. Upper spheroid is female with eggs; lower right is asexual with four gonidia that have enlarged to almost fill the entire area in the spheroid.

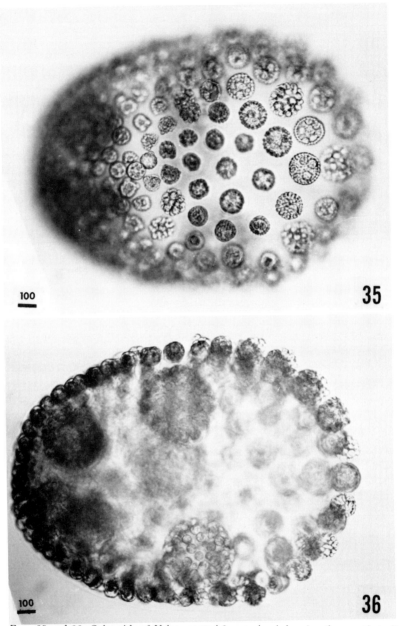

Figs. 35 and 36. Spheroids of *Volvox carteri* f. *nagariensis* bearing the genetic trait which results in the formation of fertile somatic cells.

Fig. 35. Surface view of parental spheroid showing small spheroids which have formed from fertile somatic cells. The spheroid here bears the male locus, and certain of the small spheroids arising from the somatic cells are male spheroids.

Fig. 36. Optical section of spheroid seen in Fig. 35, showing male spheroids which have been formed from gonidia inside parental spheroid.

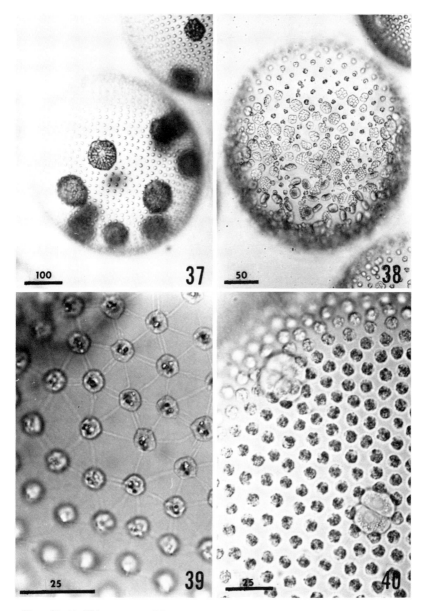

FIGS. 37-40. *Volvox aureus*, M-5 strain.

FIG. 37. Asexual spheroid with embryos at preinversion stage.

FIG. 38. Male spheroid showing large number of sperm packets forming from androgonidia in posterior two-thirds of spheroid.

FIG. 39. Somatic cells showing fine cytoplasmic connections typical of the taxonomic section Janetosphaera.

FIG. 40. Surface view of asexual spheroid showing 2-celled embryo and a 4-celled embryo. Note that cleavage has begun when gonidia are very small.

in my laboratory comparing as inocula newly released spheroids with those in which the developing spheroids were in later stages of development, results showed that, with older inoculum, values in the 80 and 90% level could be obtained; with the newly released spheroids as inoculum, the 50% level was apparently a maximum.

Darden interpreted his 50% maximum induction figure as indicative of a possible natural mechanism for preventing a population from becoming composed of 100% male individuals, a condition which would be selected against in evolution. However, even with a 100% response to inducer under laboratory conditions, this would not mean that the survival of the species in nature is necessarily in jeopardy every time the sexual response is invoked in natural populations, for it must be a rare situation in nature that one would find a synchronous population in which all individuals would be subjected to the optimal growth conditions at the same time. Spheroids under different microenvironmental conditions would ensure less than 100% response.

Induction of Parthenospores

The gonidia of the asexual spheroid in *V. aureus* may act as facultative eggs in the presence of sperm, and in certain strains (Darden, 1968, 1969; Starr, 1968) they may be converted into parthenospores by use of filtrates from a sexual male population. Parthenospores have the same appearance as mature zygotes, but having developed without fertilization, it is assumed they are haploid. The M-5 strain (Darden, 1966) responds to male filtrate by producing male embryos; the DS strain (Darden, 1968, 1969) responds to male filtrate by forming parthenosporic spheroids; the 65–98 strain (Starr, 1968) may form both males and parthenosporic spheroids. Darden (1968) showed that filtrates from sexual cultures of the M-5 strain and from the parthenosporic cultures of the DS strain were each active in inducing males in the M-5 strain and in inducing parthenospores in the DS strain. He reported that males were rarely formed in the DS strain and that the inducing factor of that strain was possibly different from the inducing factor from the M-5 strain. It is true that males are formed infrequently in the DS strain, but by examining a culture of the strain periodically during its growth period, the occasional presence of a male can be detected. In view of the potency shown by a single male of *V. carteri* f. *nagariensis*, it is not unreasonable to suspect that the inducing power of the DS strain is wholly dependent on these occasional males. Moreover, the differences in the active factors from the M-5 and the DS strains which Darden (1968) reported are better considered as differences in the biochemical processes leading to parthenospore production and male embryo production rather than in the factors themselves.

Differentiation in the Embryo

Morphological differentiation of the reproductive cells, whether the gonidia in the asexual spheroid or the androgonidia in the male, cannot be detected in the developing embryo of *V. aureus*. It is only in the enlarging young spheroid shortly before its release that one can begin to discern which type of reproductive cell it bears. Darden's observation that the developing embryo of *V. aureus* at the late preinversion stage is at the stage of maximum susceptibility is indicative that the systems in *V. aureus* and *V. carteri* are basically quite different, for the stage in the developing embryo of *V. aureus* which is most receptive to the inducer is a stage much later than that at which differentiation has already occurred in asexual and female embryos of *V. carteri* f. *nagariensis*. Addition of inducer in *V. carteri* f. *nagariensis* at the onset of cleavage has been shown to be completely ineffectual in inducing the development of the resulting embryo.

VOLVOX ROUSSELETII f. *GRIQUAENSIS* POCOCK (FIGS. 41–46)

Induction of Male and Female Spheroids

This species (Figs. 41 and 42) was the subject of the most detailed account of the morphology, development, and reproduction in any species of *Volvox* (Pocock, 1933), but as Dr. Pocock's study was confined to material from natural populations, little could be deduced as to the mechanisms of control that might be operating in determining the asexual and sexual phases. Strains of this species were isolated from collections of mud samples made near Kimberley, South Africa (Starr, 1968), and have served as material for study of an induction system (McCracken, 1969; McCracken and Starr, 1970) quite different from that found in *Volvox carteri*. In *Volvox rouseletii* we have found that, as in *V. aureus* and *V. carteri*, the inducer is produced by sexual male cultures, but the action of the inducer in *V. rousseletii* is to determine whether a reproductive cell will develop into a new embryo or a sperm packet (in the male strain) or a new embryo or an egg (in the female strain). Young spheroids of *V. rousseletii* at the time they are released from the parent contain reproductive

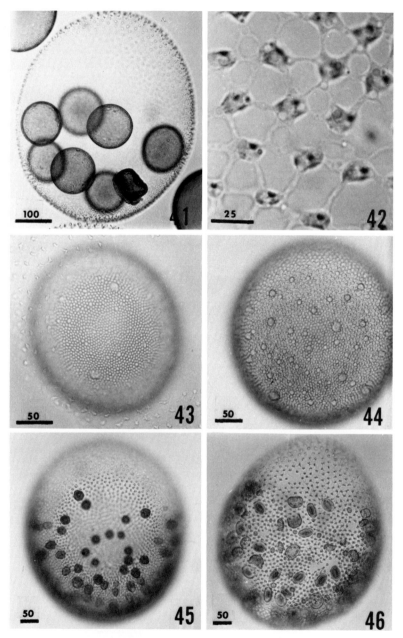

FIGS. 41–46. *Volvox rousseletii*, K-32 strains.
FIG. 41. Asexual spheroid with 7 inverted embryos and one at end of inversion.

cells of two sizes, a few much larger than the somatic cells and others intermediate between the two (Fig. 43). When inducer is not present, only the larger of the two types of reproductive cells develops and the successive cleavages result in a typical embryo (Fig. 41). When inducer is present, both large and small reproductive cells in the male strain, through a process of growth and cleavage, form sperm globoids (Fig. 46), while in the female strain the reproductive cells enlarge and are transformed directly into eggs without any cleavages ensuing (Figs. 44 and 45).

The Inducer

The inducing principle in *V. rousseletii* is believed to be proteinaceous, as its activity is markedly reduced after digestion with pronase. The raw male culture fluid can be heated at 80°C for 1 hour without reducing the activity, and temperatures of 100°C for 1 hour reduce the activity only 50%; but as we know from experiments with the inducer of *V. carteri* f. *nagariensis*, partially purified material may prove to be much more heat sensitive. Attempts at purification involved the same basic methods as were used with *V. carteri* f. *nagariensis*, i.e., adsorption on carboxymethyl cellulose, elution followed by flash evaporation, and fractionation using the Sephadex G-75. The active principle comes off in a fraction between markers, cytochrome c (MW 12,270) and bacitracin (MW 1400). Lyophilization of the most active fractions from the Sephadex G-75 column gave a dried residue with 100% activity at a concentration of 40 ng/liter. Assuming a molecular weight of 10,000, this material was active at a concentration of approximately 4×10^{-12} M.

VOLVOX GIGAS POCOCK

Strains of this species, known only from South Africa, were isolated by Starr (1968) and served as material for the study of an induction system which is similar in some respects to *V. rousseletii* but differs in others (Vande Berg and Starr, 1971). As in *V. rousseletii* the spe-

---

FIG. 42. Somatic cells showing stellate appearance resulting from broad cytoplasmic connections typical of species in the taxonomic section Euvolvox.

FIG. 43. Young spheroid just prior to release, showing large and small reproductive initials scattered among the very small somatic cells.

FIG. 44. Induced spheroid of the female strain showing enlargement of large and small reproductive cells in the formation of eggs.

FIG. 45. Female spheroid with mature eggs.

FIG. 46. Male spheroid with sperm globoids in various stages of formation. Ruptures in spheroid show that some spheroids have already matured and escaped.

cific inducer produced by the sexual male causes the reproductive initials in young spheroids to develop into sexual reproductive cells rather than function asexually to form new spheroids. As in all species of *Volvox*, the somatic cells in *V. gigas* remain small and are incapable of division when they are in either the asexual spheroid or the female spheroid. But in young spheroids of the male strain that are exposed to the inducer, not only do the large reproductive initials enlarge and cleave to form packets of sperm, but also the somatic cells in the posterior half of the male spheroid enlarge and form small sperm packets. With the exception of the mutant of *V. carteri* f. *nagariensis* discussed earlier, this is the only known species in which somatic cells can be influenced to undergo further division in the mature spheroid.

The inducer in *V. gigas* is also believed to be proteinaceous with a molecular weight around 20,000. Using Sephadex G-75 columns the fractions containing the inducer from *V. gigas* are collected after those containing the bulk of the inducer from *V. carteri* f. *nagariensis*.

OTHER SPECIES OF *VOLVOX*

Studies are now in progress in my laboratory involving the systems in other species. The doctoral investigation of Mr. Robert Karn is concerned with the induction system in *Volvox obversus*, a species which though similar to *V. carteri* in the early delimitation of reproductive cells is quite different in details of embryogenesis. The male and female strains of this species were isolated from Australia, and it was shown earlier (Starr, 1968) that the male filtrate could be used to induce the formation of sexual reproductive cells in both sexes.

Mr. David Redmon is concerned with a new variety of *Volvox dissipatrix* isolated from India. In the typical variety of this species the sexual spheroids are monoecious (producing both eggs and sperm), but in this new variety male spheroids and female spheroids are produced in separate strains. More interestingly, both the male and the female spheroids are capable of producing inducing substances which affect both sexes. In all other known induction systems in *Volvox* only the sexual male produces the inducing substance.

We now have in my laboratory clonal populations belonging to almost all the species of *Volvox* recognized by Smith in his monograph of the genus (Smith, 1944). A number of them have life cycles similar to that of *V. aureus*, in which the male spheroid is the only special sexual spheroid produced, the asexual serving as the female. To date, no induction system has been identified in *V. tertius* Meyer, *V. spermatosphaera* Powers, or *V. pocockiae* Starr. In *Volvox africanus* West, we see the ultimate in degree of variation between the sexual pattern in various clones (Starr, 1971). We have a clonal population from Missouri in which separate male and female spheroids are produced; from Australia we have clones which produce only male or female sexual spheroids, but never both; from South Africa a clone of *V. africanus* forms sexual spheroids in which the reproductive cells in a single spheroid include both sperm bundles and eggs; while from India a single clone of the species produces sexual spheroids bearing eggs and sperm together, and in addition pure male spheroids. In this Indian clone, the monoecious (eggs and sperm) spheroid and the pure male spheroid may be offspring of the same parental spheroid. No induction system has been identified in any of the clones of *V. africanus*, and nothing is known of the genetic basis for this variation among strains.

The whole problem of the control of differentiation in the monoecious species of the genus, *V. globator*, L. emend. Ehrbg., *V. barberi*, Shaw, *V. merrilli* Shaw, among others, awaits investigation.

SUMMARY

The genus *Volvox* offers a variety of species which may serve in varying ways as experimental material for studies in differentiation of a simple multicellular organism with only two kinds of cells, somatic and reproductive, the latter being asexual, male or female. In most species the differentiation of the sexual reproductive cells is in response to a specific inducer produced usually by the males, but in one case by both sexes of the species.

Volvox carteri f. *nagariensis* has been more thoroughly investigated than other species and has been shown to possess an unusual combination of characteristics which make it especially adaptable to studies of differentiation and its control. It is haploid and autotrophic, and through extended multiplication by asexual means large synchronous isogenic populations may be produced. There are separate male and female strains, both of which respond to the inducer from the male, and thus both sexual systems can be analyzed and mutants affecting the developmental process studied in the two sexes. The generation time for asexual reproduction being between 48 and 72 hours, with a 16-fold increase possible at each generation, allows for the production of a large population in a short time. Its growth in axenic culture in a defined medium permits investigations involving the uptake or release of substances in the medium, unhampered by such complex substances as coconut milk or serum needed in the culture of some other organisms. The study of genetic recombination is facilitated by the ease of zygote germination which will occur by transfer to fresh medium after a short dormancy period of approximately 3 weeks.

The development of a new individual is through synchronous successive cleavages of a uninucleate asexual reproductive cells, the gonidium, which in *V. carteri* enlarges greatly before the onset of the successive cleavages (which occur at less than hourly intervals), ultimately forming within a period of 8–10 hours an individual of 4000+ cells. It would appear that the bulk of syntheses of most cellular components has occurred prior to the onset of cleavage, but a mitotic nuclear division precedes each cleavage.

During embryogenesis in *V. carteri* f. *nagariensis* all cleavages divide the cells into two equal products except for those divisions at which the reproductive cells are delimited. In the embryogenesis of the asexual individual, the gonidia are delimited from the top 16 cells at the division of the 32-celled stage (6th cleavage); but if the gonidium from which the embryo is developing has been matured in the presence of an inducing substance from the male, embryogenesis in the female strain will result in the delimitation of egg initials from cells of the top two-thirds of the embryo at the division of the 64-celled stage (7th cleavage), while embryogenesis in the male strain will result in the delimitation of androgonidia (sperm-producing cells) at the last cleavage of the male embryo (usually the 7th, 8th, or 9th).

The inducing substance from the male is highly specific and affects only the strains of *V. carteri* f. *nagariensis*. Strains of *V. carteri* f. *weismannia* are not affected by the inducer from f. *nagariensis*, but in turn produce inducers which are specific from that form. There is even some variation in specificity among the strains of f. *weismannia*. The inducer in *V. carteri* f. *nagariensis* is believed to be proteinaceous with an estimated molecular weight of 30,000, and it is active at a concentration of 3×10^{-15} M. The site of action in the cell is unknown, but it is tempting to speculate that the inducer acts on the genome or on some immediate product, inasmuch as a mutant locus linked to the female locus eliminates the necessity of the inducing substance for the differentiation of egg-bearing spheroids.

Further mutants involving the time of differentiation, the pattern of differentiation, and the nature of the differentiated reproductive cells indicate embryogenesis in *Volvox carteri* f. *nagariensis* is under the control of a number of genetic loci and that the inducing substance is responsible only for the initiation of the sexual response.

The somatic cells in *V. carteri* show a characteristic differentiation in their loss of ability to grow and divide. This is all the more surprising inasmuch as the asexual reproductive cells, which are delimited early in embryogenesis and do not participate in many of the succeeding divisions in the embryo, later begin to enlarge and begin to cleave, while the somatic cells which have participated in every division in embryogenesis stop dividing prior to inversion of the embryo and never regain this ability. A mutant has been isolated in which the somatic cells do not lose this ability. This mutant, due apparently to a single locus, may provide us with interesting material to study controls by which such processes of cell growth and multiplication are regulated.

Induction systems in other species of *Volvox* have been described. Some systems are like that of *V. carteri* in that the inducing substance initiates a new pattern of differentiation in the developing embryo; but in systems such as are found in *V. rousseletii* and *V. gigas* the inducing substance controls the differentiation of an existing reproductive initial. In the presence of inducer the reproductive initial develops as an egg (in the female strain) or a sperm-producing cell (in the male strain); in the absence of the inducer the reproductive initial develops as a gonidium which cleaves to form the next generation. The ultimate result in all types of induction systems is the production of sexual reproductive cells, but the biochemical processes involved in the many systems must be quite varied.

Many colleagues and friends throughout the world have contributed in various ways to these investigations of *Volvox* by supplying soil samples, arranging field trips, providing work space, or by helpful discussions of various aspects of the problem. I am especially indebted to Mr. Stanley W. Anderson and Mr. Richard W. Nelson for their interest and technical assistance with all phases of the problem, to my graduate students, Mr. Warren Vande Berg and Dr. Michael McCracken, for their continuing thought-provoking discussions and cooperation, and to Dr. Harold C. Bold for his helpful criticisms of the manuscript. The work has been supported in part by grant FR-00141 from the National Institutes of Health.

REFERENCES

BARTH, L. J. (1964). "Development: Selected Topics." Addison-Wesley, Reading, Massachusetts.

COLEMAN, A. W. (1959). Sexual isolation in *Pandorina morum*. *J. Protozool.* **6**, 249–264.

DARDEN, W. H., JR. (1965). Sexual differentiation in *Volvox aureus*. Ph.D. thesis, Indiana University, Bloomington, Indiana.
DARDEN, W. H., JR. (1966). Sexual differentiation in *Volvox aureus*. *J. Protozool.* **13,** 239–255.
DARDEN, W. H., JR. (1968). Production of a male-inducing hormone by a parthenosporic *Volvox aureus*. *J. Protozool.* **15,** 412–414.
DARDEN, W. H., JR., and SAYERS, E. R. (1969). Parthenospore induction in *Volvox aureus* DS. *Microbios* **2,** 171–176.
DOBELL, C. (1932). "Antony van Leeuwenhoek and His "Little Animals." Staples Press, London.
GERISCH, G. (1959). Die Zelldifferenzierung bei *Pleodorina californica* Shaw und die Organisation der Phytomonadinenkolonien. *Arch. Protistenk.* **104,** 292–358.
GOLDSTEIN, M. (1964). Speciation and mating behavior in *Eudorina*. *J. Protozool.* **11,** 317–344.
JANET, C. (1923). "Le *Volvox*. Troisième Mémoire." Protat Frères. Macon.
KOCHERT, G. (1968). Differentiation of reproductive cells in *Volvox carteri*. *J. Protozool.* **15,** 438–452.
KOFOID, C. A. (1898). Plankton studies. II. On *Pleodorina illinoisensis*, a new species from the plankton of the Illinois River. *Bull. Ill. State Lab. Nat. Hist.* **5,** 273–293.
McCRACKEN, M. D. (1969). Induction and development of reproductive cells in the K-32 strains of *Volvox rousseletii*. Ph.D. thesis, Indiana University, Bloomington, Indiana.
McCRACKEN, M.D., and STARR, R. C. (1970). Induction and development of reproductive cells in the K-32 strains of *Volvox rousseletii*. *Arch. Protistenk.* **112,** 262–282.
POCOCK, M. A. (1933). *Volvox* in South Africa. *Ann. South Afr. Mus.* **16,** 523–646.
POWERS, J. H. (1907). New forms of *Volvox*. *Trans. Amer. Microsc. Soc.* **27,** 123–149.
POWERS, J. H. (1908). Further studies in *Volvox*, with descriptions of three new species. *Trans. Amer. Microsc. Soc.* **28,** 141–175.
PROVASOLI, L., and PINTNER, I. J. (1959). Artificial media for freshwater algae; problems and suggestions. *In* "The Ecology of Algae" (C. A. Tryon and R. T. Hartman, eds.), pp. 84–96. Spec. Publ. No. 2, Pymatuning Lab. of Field Biology, University of Pittsburgh.
SHAW, W. R. (1894). *Pleodorina*, a new genus of the Volvocineae. *Bot. Gaz.* **19,** 279–283.
SMITH, G. M. (1944). A comparative study of the species of *Volvox*. *Trans. Amer. Microsc. Soc.* **63,** 265–310.
STARR, R. C. (1968). Cellular differentiation in *Volvox*. *Proc. Nat. Acad. Sci. U. S.* **49,** 1082–1088.
STARR, R. C. (1969). Structure, reproduction, and differentiation in *Volvox carteri* f. *nagariensis* Iyengar, strains Hk 9 & 10. *Arch. Protistenk.* **111,** 204–222.
STARR, R. C. (1971). Sexual reproduction in *Volvox africanus*. "Contributions to Phycology" (B. C. Parker and R. M. Brown, eds.). In press.
STEIN, J. R. (1958). A morphologic and genetic study of *Gonium pectorale*. *Amer. J. Bot.* **45,** 664–672.
VANDE BERG, W. J., and STARR, R. C. (1971). Structure, reproduction and differentiation in *Volvox gigas* and *Volvox powersii*. *Arch. Protistenk.* in press.

On the Development of Insulin Sensitivity by Mouse Mammary Gland in Vitro

YALE J. TOPPER, S. H. FRIEDBERG, AND TAKAMI OKA

National Institute of Arthritis and Metabolic Diseases, National Institutes of Health, Bethesda, Maryland 20014

INTRODUCTION

The epithelial cells in the mammary gland of the mature, nonpregnant animal undergo virtually no proliferation. In contrast, the cells in the gland of the pregnant animal undergo extensive proliferation. One of the important questions concerning the development of this gland relates to the mechanisms by which epithelial cell multiplication is prohibited in the mature nonpregnant animal, and initiated at the start of pregnancy. It has been shown (Stockdale and Topper, 1966) that mammary gland explants from mice in mid-pregnancy respond rapidly to insulin in terms of increased epithelial DNA synthesis. Explants from the glands of mature nonpregnant mice also respond to insulin in this respect, but do so only after a considerable lag period (Stockdale and Topper, 1966). This lag period has now been further studied, and it will be discussed in the context of the prohibition and initiation of DNA synthesis referred to above. It will also be considered in relation to the following question. If the initiation, by exogenous insulin, of DNA synthesis in virgin explants has a physiological counterpart, why does the endogenous hormone not promote DNA synthesis in the mammary epithelial cells of the intact, mature virgin animal?

Another intriguing question relating to the development of the mammary gland concerns the formation of rough endoplasmic reticulum (RER). RER is necessary for the production of secretory proteins, but is not necessary for the production of nonsecretory proteins by mammary gland and other tissues. Virtually none of the mammary epithelial cells in the virgin animal contains appreciable RER, some of the cells in the mid-pregnant animal have extensive RER, but all lactating cells have a rich supply of these organelles. Glucocorticoids, such as hydrocortisone, in the presence of insulin, have been shown (Mills and Topper, 1969) to promote the formation of these membranes in explants derived from the pregnant mouse. It was subsequently demonstrated (Hollmann, unpublished) that the RER content of lactating cells in the intact animal could be increased by administration of hydrocortisone. Why do endogenous glucocorticoids, in conjunction with endogenous insulin, not promote formation of RER in mammary cells of the nonpregnant animal?

Certain more general aspects of hormone action, as they relate to problems of development, will also be considered.

MATERIALS AND METHODS

The previously described (Juergens *et al.*, 1965) organ culture technique employing mammary gland explants was used. Such explants contain epithelial and fat cells, predominantly. In order to ascertain whether or not the observed effects related to epithelial cells, the tissue was treated with collagenase and the epithelial cells were separated from the lysed fat cells by centrifugation (Lasfargues, 1957). In some instances this separation was performed after the explants had been cultured. In other cases fat cells were removed initially, and the residual epithelial cell–collagen complex was then cultured using the same technique employed with explants. By these means it has been determined that the results herein reported reflect responses of the epithelial cells.

Insulin (crystalline beef insulin, a gift from the Eli Lilly Co.) was used at a calculated final concentration of 10^{-7} M. However, prevention of hormone loss, consequent to adsorption on glassware, by addition of bovine serum albumin (final concentration, 2.5%) permitted the same effects to be manifested at a calculated insulin concentration of 10^{-9} M. Hydrocortisone was used at a final concentration of 10^{-5} M, but we have previously reported that the minimal effective concentration of this hormone is 10^{-8} M. Medium 199 (Microbiological Associates) containing fructose instead of glucose was used in all experiments.

RESULTS

Since we had observed (Stockdale and Topper, 1966) a lag in the response of virgin mammary epithelium to insulin in terms of DNA synthesis, it was pertinent to determine whether there is a lag in terms of other parameters. In essence, then, this report will compare the time courses of several responses to insulin of the epithelial cells in pregnancy and virgin tissues.

Figure 1 shows the initial time course of accumulation of the nonmetabolizable amino acid, α-aminoisobutyric acid (AIB) by pregnancy and virgin explants in the presence and absence of insulin. It can be

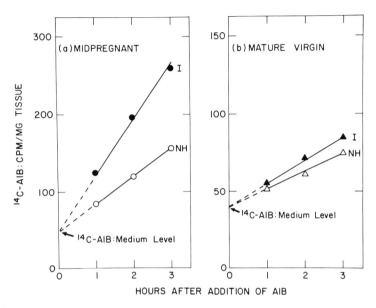

FIG. 1. Effect of insulin on the accumulation of α-aminoisobutyric acid-1-^{14}C (^{14}C-AIB) by mammary explants from mid-pregnant (a) and 4-month-old virgin (b) mice. The explants were incubated in the absence (NH) or in the presence (I) of insulin. After 1 hour of incubation ^{14}C-AIB (sp. act. 6.49 mCi/mmole) was added to each culture dish to a final level of 0.1 μCi/ml. At the indicated times, accumulation of ^{14}C-AIB was measured as follows. The explants were weighed, placed on glass filter paper, and washed 3 times with Medium 199 and 2 times with water by suction filtration. The explants and filter papers were transferred to scintillation vials and dissolved in 1 ml of NCS at 50° overnight. Ten milliliters of scintillation fluid (4 gm POP, 0.4 gm POPOP per liter of toluene) was added to each vial, and ^{14}C was measured in a liquid scintillation spectrometer. Each point represents the average of 3 determinations.

seen that for the first 3 hours after the addition of AIB the rates are linear, and that during this time insulin exerts a much greater effect on the pregnancy than on the virgin explants. The hormone had no effect on the inulin space in either tissue.

In the next series of experiments (Fig. 2) the rate of accumulation of AIB at various time points during 96 hours of culture was followed. The rate of accumulation of the amino acid by pregnancy explants is constant over the entire 4-day culture period in the presence and absence of insulin. The increment in accumulation between these two systems is, therefore, also constant during this time. The rate of accumulation by virgin explants in the absence of insulin was also constant during 4 days. During the first day the rate of accumulation by virgin tissue

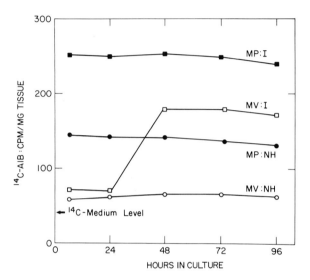

FIG. 2. The accumulation of α-aminoisobutyric acid-1-^{14}C (^{14}C-AIB) during 96 hours of culture. Mammary gland explants prepared from mid-pregnant (MP) and 4-month-old virgin (MV) mice were cultured in the absence (NH) or in the presence (I) of insulin. ^{14}C-AIB (0.1 µCi/ml) was present for 3 hours ending at the times indicated, and accumulation was measured as described in the legend to Fig. 1. Each point represents the average of 3 determinations.

in the presence of the hormone was constant, and it was only slightly larger than in the absence of insulin. Thus, again we see a relatively small response initially. By the end of the second day, however, insulin exerted a marked effect on the rate in the virgin explants. This enhanced rate of accumulation was maintained during the next 2 days. The pattern presented in Fig. 2 was also observed when corresponding epithelial cell–collagen complexes were used instead of explants. This indicates that the effects reported reflect responses of the epithelial cells, and that the development of insulin responsiveness in terms of this parameter does not require a contribution from the fat cells.

The intent of the studies shown in Fig. 3 was to delineate more precisely the duration of the lag period in the virgin tissue. The lag period lasts 24–28 hours.

The experiments presented in Fig. 4 were designed to determine whether incubation of virgin tissue in the absence of insulin for varying lengths of time had an effect on the kinetics of response to the hormone added at later time points. It can be seen that preincubation for 12 or 24 hours in the absence of insulin had essentially no effect on these kinetics. The duration of the lag period and the events which lead from hormone-insensitivity to sensitivity are, therefore, independent of exogenous insulin. Moreover, once the tissue has traversed the insensitive period in the absence of the hormone, the response to added insulin is very rapid (NH_{32}–I_{32-36} system).

It was stated in the introduction that the formation of RER by alveolar cells in pregnancy explants requires both insulin and glucocorticoid. The results of experiments reported in Table 1 represent an attempt to understand why endogenous insulin and glucocorticoid do not promote RER in the mammary epithelial cells of the intact mature, nonpregnant mouse. In these studies the membrane-associated enzyme, NADH-cytochrome c reductase, in epithelial cells of pregnancy and virgin explants, was assayed after 48 and 96 hours of culture under the conditions shown. The data reveal that insulin and hydrocortisone alone effected small increases in the enzyme activity in pregnancy tissue after 48 hours, but that the effect was much

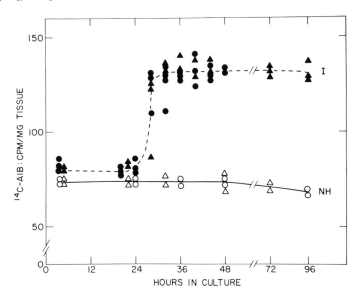

FIG. 3. Time course of AIB-accumulation by mammary explants derived from mature virgin (4-month-old) mice. Explants from 2 mice were cultured separately in the absence (NH) or in the presence (I) of insulin. The accumulation of AIB was measured after pulsing with ^{14}C-AIB (0.1 μCi/ml) for 3 hours ending at the times indicated. Each point represents a single determination. Circles represent values from one animal; and triangles, values from the other.

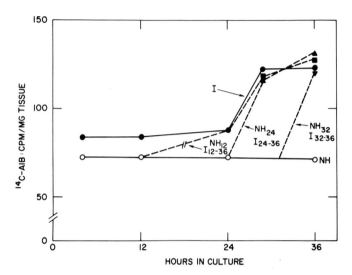

FIG. 4. Effect of culture in absence of insulin on the subsequent insulin-stimulated accumulation of AIB by virgin explants. During culture in the absence (NH) of insulin, virgin explants were transferred to insulin (I) medium at the indicated times. AIB-accumulation was measured after pulsing with ^{14}C-AIB (0.1 μCi/ml) for 3 hours ending at the times shown. Each point represents the average of 3 separate determinations.

greater when both hormones were present. Mature virgin tissue responded to each hormone alone to about the same extent as pregnancy tissue after 48 hours, but the combination of the two produced no greater effect at this time. However, after 96 hours the double hormone system did elicit a much greater stimulation than the single hormone systems. These results, together with those relating to AIB uptake and preliminary observations (Oka and Topper, unpublished) on hydrocortisone-mediated increases in the rate of choline incorporation into membrane phospholipid, suggest that the small effect seen with the combination of the two hormones in the virgin system after 48 hours is due to initial insensitivity to insulin rather than to hydrocortisone. Note that sensitivity to insulin in terms of NADH–cytochrome c reductase develops later than it does with respect to AIB uptake. A possible inference from these results is that RER is not formed in the mammary epithelial cells within the intact virgin animal not because of insensitivity to glucocorticoid, but for lack of sensitivity to insulin.

Glucose-6-phosphate dehydrogenase (G-6-P-D) is an intracellular protein present in the soluble fraction of the cytoplasm. Its formation in mammary gland has been shown to be dependent upon insulin (Leader and Barry, 1969). Figure 5 compares the time courses of the combined activities of G-6-P-D and gluconate-6-phosphate dehydrogenase found in pregnancy and virgin tissue during culture in the presence of insulin. In each instance, the soluble fraction of the epithelial cell pellet obtained by collagenase treatment of cultured explants was used for assay. Again it is seen that the response of the virgin tissue to the hormone occurs 1–2 days later than the response of the pregnancy tissue. With virgin tissue the enzyme activity decreased progressively during culture in the absence of insulin; no increase occurred in pregnancy tissue in the absence of the hormone.

Results of experiments bearing on the question as to why endogenous insulin does not promote DNA synthesis in mammary epithelial cells of the intact, mature virgin animal, whereas exogenous insulin does initiate DNA synthesis in the cells of the explanted tissue will be presented next. The rate of incorporation of tritiated thymidine into trichloroacetic acid-insoluble material within the explants has been used as a reflection of DNA synthesis by the epithelial cells. This is based on autoradiographic studies (Stockdale et al., 1966) which demonstrated silver grains over epithelial nuclei, but not over fat cell nuclei, and on the observed correspondence between the time courses of thymidine-^{3}H incorporation and mitotic indices (Stockdale and Topper, 1966).

TABLE 1
Changes in NADH-Cytochrome c Reductase[a]

Conditions	Mature virgin		Midpregnant 48 Hours[b] (%)
	48 Hours[b] (%)[c]	96 Hours[b] (%)	
No hormone	85	105	95
Insulin	100	135	125
Hydrocortisone	130	176	138
Hydrocortisone + insulin	130	420	310

[a] Forty-sixty milligrams of mammary explants was prepared from mid-pregnant or mature virgin mice for each hormone system. At the end of culture the explants from each system were weighed and treated with collagenase (Lasfargues, 1957). The resulting epithelial pellets were homogenized in 2 ml of cold 0.25 M sucrose, and the homogenates were centrifuged at 5500 g for 20 minutes at 4°C. The supernatant was assayed for NADH-cytochrome c reductase activity (Ernster et al., 1962).

[b] Culture period.

[c] Values are expressed as percentage of zero time control.

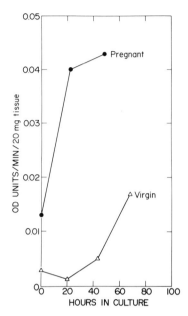

Fig. 5. Levels of the combined activities of glucose-6-phosphate dehydrogenase and 6-phosphogluconate dehydrogenase as a function of time. Mammary explants from mid-pregnant and mature virgin mice were cultured in the presence of insulin. At the indicated times the tissue was treated with collagenase (Lasfargues, 1957), and extracts from the epithelial pellets were assayed for the enzymes (Glock and McLean, 1954).

Figure 6 shows that DNA synthesis by pregnancy explants is relatively active during the first 6 hours in the presence (I) and absence (NH) of insulin, and that it declines more slowly in the next 6 hours in the I than in the NH system. Whereas the NH system continues to decline, the I system plateaus between 12 and 18 hours, and then rapidly increases to a maximum (Stockdale and Topper, 1966) (not shown) between 18 and 24 hours. The very low initial rate of DNA synthesis by virgin explants is maintained during the first 24 hours in both the I and the NH systems. The rate in the NH system increases little during the second day, but the rate in the I medium increases markedly after 36 hours. Pregnancy tissue, then, responds to insulin in terms of DNA synthesis at least 18 hours before virgin tissue, and the corresponding maximum rates occur about 24 hours apart (Stockdale and Topper, 1966).

The results of experiments in which virgin tissue was cultured

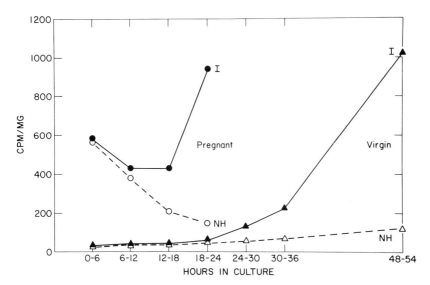

FIG. 6. DNA synthesis in mammary gland explants from pregnant and virgin mice. Incubations were carried out in the absence (NH) or presence (I) of insulin, and thymidinemethyl-^{3}H (sp. act. 20 Ci/mmole) at a final level of 1 μCi/ml was added for 6-hour periods during the culture as indicated. The incorporation of isotope into DNA was determined by a modification of a method described previously (Stockdale and Topper, 1966). Each point represents the average of 3 determinations.

either with insulin throughout the entire culture period, or initially in the absence of insulin for varying periods, are presented in Fig. 7. The rates of DNA synthesis are about the same in all the systems for the first 24 hours. Between 24 and 30 hours the rates in the NH_6I, $NH_{12}I$, and $NH_{18}I$ systems are the same, but a little slower than that in the I system. However, between 24–36 hours the slopes in all cases are essentially identical. It appears that whereas virgin tissue may be slightly sensitive to insulin during the first 18 hours as reflected by DNA systhesis, a situation reminiscent of the AIB parameter, an important insulin-independent component is implicated in the eventual acquisition of greatly enhanced responsiveness to the hormone.

Between 30 and 36 hours the $NH_{24}I$ system has a much slower rate than the others discussed above, but, as shown in Fig. 8, it can attain the same maximum at the same time as the I system. Note that the NH system also reaches its maximum, albeit a much lower one, at 48 hours. This demonstration that in isolated virgin tissue the time at which peak DNA synthesis occurs is not a function of the temporal

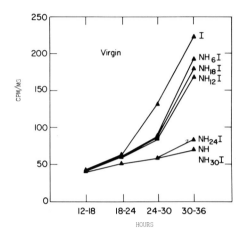

FIG. 7. Delay between the onset of insulin treatment and the DNA synthetic response. Virgin mammary gland tissue was cultured for varying lengths of time in hormone-free (NH) medium before exposure to insulin (I). Cultures were pulsed with thymidinemethyl-^{3}H and assayed as described in the legend to Fig. 6.

aspect of addition of insulin fortifies the conclusion that the lag period in such tissue is, indeed, independent of exogenous insulin.

Table 2 provides a summary of the time elements required for manifestation of various stimulatory effects of insulin on mid-pregnancy and virgin mouse mammary explants.

DISCUSSION

It has been shown that pregnancy mammary epithelial cells from the mouse are sensitive to insulin initially *in vitro* in terms of several parameters, but that freshly explanted virgin tissue is either insensitive or only slightly sensitive. After one to several days in culture the virgin tissue acquires sensitivity to the hormone comparable to that of the pregnancy tissue. Acquisition of the ability to respond to insulin is largely independent of exogenous insulin. Differences in the length of lag period as a function of particular responses indicate that we are probably dealing with mechanistically different manifestations of action of the hormone. The events which lead to insulin sensitivity *in vivo* during pregnancy are probably not identical to those which produce sensitivity in virgin tissue *in vitro*. Thus, altered levels of some hormones such as estrogen and progesterone and prolactin, and the *de novo* appearance of others, such as placental lactogen, may play a role in the intact animal during pregnancy.

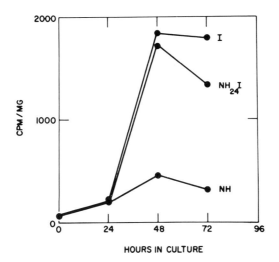

FIG. 8. Time courses of DNA synthesis by virgin mammary explants cultured for 72 hours in the absence (NH) of insulin, and in the presence (I) of insulin, or in the absence of insulin for 24 hours followed by exposure to insulin for 48 hours. Cultures were pulsed with thymidinemethyl-^{3}H and assayed as described in the legend to Fig. 6.

TABLE 2
RESPONSE TO INSULIN. COMPARISON OF LAG PERIODS FOR VARIOUS PARAMETERS IN MID-PREGNANCY AND VIRGIN EXPLANTS

Tissue	AIB uptake (hours)	NADH–cytochrome c reductase (hours)	G-6-P-D (hours)	DNA synthesis (hours)
Mid-pregnancy	≤1	<48	<24	<12
Mature virgin	24–28	48–96	24–48	24–30

These factors clearly do not participate in the conversion of insulin-insensitive into sensitive virgin explants under the conditions employed. This conversion seems to be a consequence, particularly, of removal of the tissue from the animal. It is worth noting, in this context, that transplantation of nongrowing ducts from the glands of mature, nonpregnant mice into epithelium-free mammary fat pads of comparable animals results in extensive proliferation of the ducts.

If insulin is at least partially responsible for the initiation of epithelial DNA synthesis during pregnancy, the results could explain why this hormone does not perform a similar function in the mature, nonpregnant animal. That is, for some as yet unknown reason the potentially sensitive tissue is, in fact, insensitive to insulin in this physiological state. Tissue isolation or pregnancy can confer insulin sensitivity to the gland.

These observations also have more general implications for hormone interactions with tissues. If a given hormone has more than one target tissue, and each tissue is maximally sensitive at all times, modulations in the plasma level of the hormone would not be expected to lead to highly selective effects on a particular target. The selectivity of a hormone response could be enhanced if the sensitivity of one or another target tissue varied under different physiological conditions. Furthermore, if this were true the plasma level of hormones would not necessarily be an accurate reflection of hormone-tissue interaction.

Changes in sensitivity to various stimuli, such as hormones, also appear to impinge on fundamental problems of developmental biology. Responsiveness may be acquired or lost during maturation. Isolated, human fetal islets of Langerhans secrete insulin in response to glucagon, but not in response to glucose or tolbutamide. Sensitivity to the latter stimuli develops later (Espinosa de los Monteros M. et al., 1970). Arginase activity in the liver of late fetal rats is increased by the administration of thyroxine, but not by hydrocortisone. The liver of the 5-day postnatal rat, on the other hand, no longer responds to thyroxine in this way, but does so following injection of hydrocortisone (Greengard et al., 1970). The isolated tissue system described in this report may provide a useful model for studying such phenomena.

REFERENCES

ERNSTER, L., SIEKEVITZ, P., and PALADE, G. E. (1962). Enzyme-structure relationships in the endoplasmic reticulum of rat liver. A morphological and biochemical study. J. Cell Biol. **15,** 541–562.

ESPINOSA DE LOS MONTEROS M., A., DRISCOLL, S. G., and STEINKE, J. (1970). Insulin release from isolated human fetal pancreatic islets. Science **168,** 1111–1112.

GLOCK, G. E., and McLEAN, P. (1954). Levels of enzymes of the direct oxidative pathway of carbohydrate metabolism in mammalian tissues and tumors. Biochem. J. **56,** 171–175.

GREENGARD, O., SAHIB, M. K., and KNOX, W. E. (1970). Developmental formation and distribution of arginase in rat tissues. Arch. Biochem. Biophys. **137,** 477–482.

HOLLMANN, K. H. Unpublished observations.

JUERGENS, W. G., STOCKDALE, F. E., TOPPER, Y. J., and ELIAS, J. J. (1965). Hormone-dependent differentiation of mammary gland in vitro. Proc. Nat. Acad. Sci. U.S. **54,** 629–634.

LASFARGUES, E. Y. (1957). Cultivation and behavior *in vitro* of the normal mammary epithelium of the adult mouse. *Anat. Rec.* **127,** 117–129.

LEADER, D. P., and BARRY, J. M. (1969). Increase in activity of glucose-6-phosphate dehydrogenase in mouse mammary tissue cultured with insulin. *Biochem. J.* **113,** 175–182.

MILLS, E. S., and TOPPER, Y. J. (1969). Mammary alveolar epithelial cells: Effect of hydrocortisone on ultrastructure. *Science* **165,** 1127–1128.

OKA, T., and TOPPER, Y. J. Unpublished observations.

STOCKDALE, F. E., and TOPPER, Y. J. (1966). The role of DNA synthesis and mitosis in hormone-dependent differentiation. *Proc. Nat. Acad. Sci. U.S.* **56,** 1283–1289.

STOCKDALE, F. E., JUERGENS, W. G., and TOPPER, Y. J. (1966). A histological and biochemical study of hormone-dependent differentiation of mammary gland tissue *in vitro*. *Develop. Biol.* **13,** 266–281.

Development of Hormone-Responsive Cell Strains in Vitro[1]

MARTIN POSNER,[2] WILLIAM J. GARTLAND, JEFFREY L. CLARK, GORDON SATO, AND CARL A. HIRSCH

Department of Biological Sciences, University of California, San Diego, California 92037 and Department of Medicine, Beth Israel Hospital, Boston, Massachusetts 02215

INTRODUCTION

Previous work in this laboratory has demonstrated the feasibility of selecting for permanent mammalian cell strains in culture which retain the differentiated properties of the tissue from which they were derived (Buonassisi et al., 1962).

A number of such cell strains have been isolated from functional endocrine tumors in animals by the use of a serial technique of alternate culture *in vitro* and transplantation back into animals of a syngeneic strain. Using this method, selective enrichment of the tumor is achieved for epithelial cells which can adapt and grow in tissue culture and which can then be isolated from connective tissue fibroblast cells by standard cloning techniques.

By such techniques a number of clonally derived, permanently established, differentiated cell strains have been obtained. These strains include (1) ACTH responsive, steroid-secreting adrenal cortex cells (Yasumura et al., 1966a), (2) ACTH-secreting pituitary cells (Yasumura and Sato, 1967), (3) growth hormone-secreting pituitary cells (Yasumura et al., 1966b; Tashjian et al., 1968), (4) cyclic AMP-responsive, steroid-secreting, mouse Leydig cells (Shin, 1967), (5) hormone-secreting rat Leydig cells (Shin et al., 1968), (6) glucocorticoid-responsive rat glial cells (Benda et al., 1968), (7) neuroblastoma cells capable of synthesizing epinephrine and acetylcholine (Augusti-Tocco and Sato, 1969), and (8) teratoma cells which maintain their multipotent capacity for differentiation (Rosenthal et al., 1970).

---

[1] Supported by a Scholar's Grant from the American Cancer Society (PS No. 50).

[2] This work was supported by the research grants from the National Institutes of Health (CA 04123) and National Science Foundation (GB-7104, GB-81641).

SELECTION FOR HORMONE-DEPENDENT CELL STRAINS

Recently, a systematic program has been started in our laboratory to select for permanent cell strains which not only maintain differentiated cell function in culture, but whose growth is also hormone dependent or hormone responsive.

The technique used for obtaining such cell strains is a modification of that previously described and involves a systematic effort to select for those cells in the culture whose growth is hormone dependent and elimination of all other cells from the culture.

If one starts with an animal tumor whose growth is hormone dependent in the animal, one cultures the tumor *in vitro* in the presence of the stimulatory hormone, or, if the effect of the hormone is inhibitory, in its total absence. At selected times during culture of the tumor the hormonal environment of the cells is altered so that only hormone-*independent* cells can grow (e.g., removal of stimulatory hormone) and 5-bromodeoxyuridine is added to the culture medium in the absence of thymidine. Cells whose growth is hormone independent will continue to grow and incorporate 5-bromodeoxyuridine into their DNA and will be killed when the culture is subsequently exposed to visible light on the blue side of the visible spectrum (e.g., fluorescent lamps, Puck and Kao, 1967).

By means of such a technique, connective tissue fibroblasts and hormone-independent epithelial cells can be eliminated from the cell culture and under favorable conditions clonal lines of cells can be directly obtained from a first culture of the tumor without making multiple passages back through the animal. If multiple passage back through the animal is necessary, the animals must be primed to have high levels of circulating stimulatory hormones, or endocrinectomized so that inhibitory hormones are absent.

With such techniques, a line of mammary tumor cells has been isolated whose growth is dependent on the tropic pituitary hormones, prolactin and growth hormone, plus insulin and is described in detail in the following section. Work is also currently proceeding to isolate a line of mouse pituitary cells which make thyroid-stimulating hormone (TSH) and whose growth is inhibited by thyroxine. This line of cells is derived from a mouse pituitary tumor of similar properties *in vivo* and was induced by thyroidectomy (Dent et al., 1955, 1956).

Attempts are also being made to induce hormone-dependent tumors in animals suitably primed with stimulatory hormones. One method is to use chemical carcinogens in a subcarcinogenic dose and to simultaneously graft a pituitary tumor making large quantities of stimulatory hormone for the target tissue (e.g., mammary gland, Furth and Kim, 1960c). In the case of the gonads and adrenals, castrations and adrenalectomies are performed and the tissue is minced and transplanted back into the spleen of the same animal. This should result in high production of tropic stimulatory hormone by the pituitary of the animal (FSH or ACTH), since the spleen is drained by the hepatic portal system into the liver, which inactivates steroid hormones (Furth and Sobel, 1947). In principle, this method can be coupled with other carcinogens to induce hormone-dependent tumors, e.g., X-ray irradiation.

DEVELOPMENT OF A HORMONE-DEPENDENT MAMMARY TUMOR CELL LINE

With the selection technique described in the previous section a hormone-dependent mammary tumor cell line has been isolated whose growth in culture is dependent on the presence of rat prolactin and rat growth hormone, plus insulin. This cell line was derived from hormone-responsive tumors induced in female Wistar-Furth rats by gastric ingestion of repeated dosages of 3-methylcholanthrene, (Furth and Kim, 1960a, b).

Epithelial cells from the original tumor grew well enough in the first passage in culture to compete with connective tissue fibroblasts, which could be eliminated by means of the bromodeoxyuridine treatment in the absence of hormones, as described in the previous section. The bromodeoxyuridine treatment consisted of leaving the cells in thymidineless medium with 10 μg of bromodeoxyuridine per milliter for 48 hours (approximately 1.5–2 doubling times) followed by a 1 hour exposure to light from a fluorescent lamp.

The isolated cell line grew with a doubling time of 24–36 hours under optimum conditions of maximum hormonal stimulation. This hormonal stimulation could be obtained in three different ways: (1) a combination of insulin plus rat prolaction and rat growth hormone, (2) insulin plus 10–15% conditioned medium from a line of rat pituitary cells making rat prolactin and rat growth hormone (the GH_3 pituitary line, Tashjian et al., 1968), (3) use of fetal calf serum (2.5%), which is apparently rich in lactogenic hormones. Neither bovine nor ovine prolactin and growth hormone has proved effective in stimulating cell growth, nor have estrogenic hormones.

The basic medium used for most experiments was either Ham's F-10 or Dulbecco Modified Eagle's Medium (DME) with 10% hypophysectomized dog serum. Gelding serum was also tried, but it appears to support growth of the cells after an initial period of adaptation, presumably because of the presence of growth hormone and residual levels of prolactin.

EXPERIMENTS USING HORMONE-DEPENDENT CELL STRAINS

A number of growth experiments with hormones have been done on these mammary cells; replicate culture flasks plated with equal numbers of cells were used. Some of these experiments were done with cells selected for hormone dependence using the bromodeoxyuridine treatment as described, but otherwise unselected (i.e., not clonally selected). The later experiments (e.g., insulin dilution experiments, Fig. 3) were done with cell lines that had been clonally selected by single-cell plating techniques, as well as selected for hormone dependence by bromodeoxyuridine treatments.

Figure 1 shows a growth curve for uncloned mammary cells previously selected for hormone dependence with BUDR. The experiment was done using Ham's F-10 medium with hypophysectomized dog serum (10%). Rat prolactin and growth hormone, and insulin were added to half the flasks as indicated, and the control flasks were maintained in the basic medium without hormones. As shown in Fig. 1 the cells in the flasks with hormones started to grow within 2–3 days, whereas the cells in the control flasks showed no significant growth for at least a week. However, after the first week cells in the control flasks without hormones appeared to grow at almost the same rate as the cells in the flasks receiving hormones. A possible explanation for this is that the bromodeoxyuridine treatment was not 100% successful and a small fraction of the cells remaining were still hormone independent, whereas the majority were hormone dependent. According to this explanation, the apparent "lag" period of a week in the control flasks without hormones represents the time needed for the hormone-independent fraction of the cells to reach the same number as the hormone-dependent fraction, and thus, by extrapolating the growth phase of the curve (second week on) back to zero time, one obtains the number and fraction of hormone-independent cells in the initial cell culture (approximately 20%).

Figure 2 shows the results of an insulin dilution experiment with and without rat prolactin and rat growth hormone. All flasks were grown in a basic medium of Dulbecco's Modified Eagle's Medium

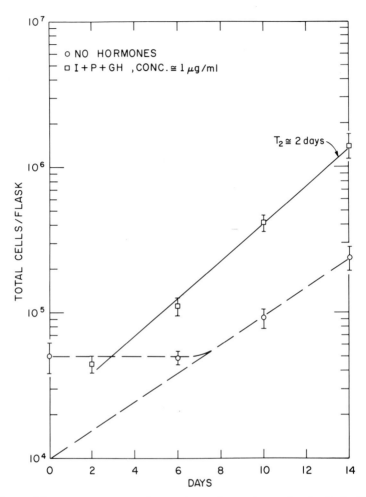

Fig. 1. Mammary tumor growth experiment done using uncloned cells in Ham's F-10 culture medium with 10% hypophysectomized dog serum. The rat prolactin and rat growth hormones used in the experiment were obtained from the National Institute of Arthritis and Metabolic Diseases, NIH, Bethesda, Maryland.

(DME) plus 10% hypophysectomized dog serum. Hormones were added to some of the flasks as indicated, and all flasks were allowed to grow for 20 days and then the number of cells were counted using a hemacytometer. The cells used in this experiment were a *clonal* line which had previously undergone a bromodeoxyuridine treatment to select for hormone dependence.

In this experiment purified bovine insulin was used, and the insulin

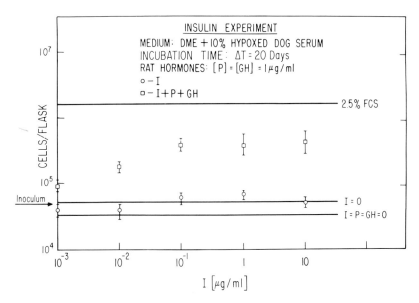

FIG. 2. Insulin dilution experiment done using Dulbecco modified Eagle's Medium (DME) with 10% hypophysectomized dog serum and purified bovine insulin, and cloned mammary cells.

concentration was varied in decade steps, while all other hormone concentrations remained fixed. One positive control and two negative controls were also run. The positive control consisted of growing the cells in the basic medium with no hormones but with 2.5% fetal calf serum, which appears to contain all the hormones necessary for the growth of the cells. The negative controls consisted of one flask with no hormones added of any kind (I=P=GH=0) and one flask with rat prolactin and growth hormone, but no insulin (I=0). The number of cells in the control flasks at the end of the growth period is indicated by the appropriately labeled lines.

The main part of the experiment thus consists of observing the growth during the incubation period (20 days) as a function of insulin concentration, with and without rat prolactin and growth hormone (squares and circles). The initial inoculum of cells was of the order of 5×10^4 cells per flask and is indicated by the arrow on the left-hand margin of Fig. 2.

The results of the insulin dilution experiment shown in Fig. 3 can be summarized as follows:

1. There is no significant growth in the cell cultures without hor-

mones (i.e., the cell number does not increase, see lowest horizontal line in Fig. 2), and very little growth in the absence of insulin (next to lowest horizontal line in Fig. 2) even if both rat prolactin and growth hormone are present.

2. With insulin plus rat prolactin and rat growth hormone present, significant growth does occur, although even at the highest insulin concentrations used the growth of the cells does not equal that of the cells in the fetal calf serum control flasks. Recent ex-

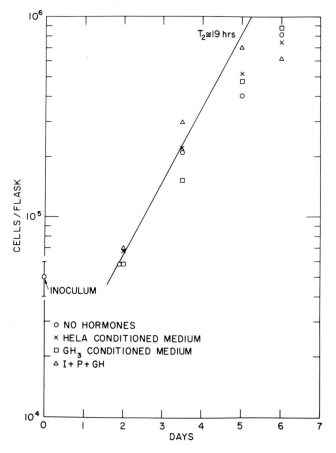

FIG. 3. HeLa growth experiment with and without lactogenic hormones. The culture medium used was Ham's F-10 with 10% hypophysectomized dog serum. The conditioned media used were harvested from large culture flasks of HeLa and GH_3 pituitary cell cultures after exposure to the cells for periods of 4–7 days.

periments indicate that this may be due to adrenal corticosteroid hormones present in fetal calf serum that are not present in hypophysectomized animal serum.

3. The insulin concentration at which cell growth saturates (0.1 micrograms/ml) is at least a factor of 10 less than that obtained by Turkington for the stimulation of normal mouse mammary gland *in vitro* (Turkington, 1968); a possible reason for this difference is our use of purified bovine insulin (obtained from Professor Steiner at the University of Chicago). When similar experiments were tried with clinical grade insulin from a local pharmacy (Iletin), a higher concentration of insulin was necessary to reach a maximum rate of cell growth.

Similar experiments to the insulin dilution experiment shown in Fig. 3 have been done recently with proinsulin; they indicate that proinsulin will also support mammary cell growth in combination with rat prolactin and/or rat growth hormone, but at a higher concentration (0.1–1.0 μg/ml of proinsulin).

The fact that the cloned cells used for the insulin dilution experiment showed no significant growth in the absence of hormones (lowest line in Fig. 2) indicates that they were indeed hormone dependent rather than hormone responsive. This verifies the hypothesis used to explain the results of the growth experiment with unCloned cells shown in Fig. 1, i.e., that the cells contained a mixture of hormone-dependent and hormone-independent cells, and that the hormone-independent fraction (of the order of 20% of the total) could be removed by single-cell cloning techniques.

Experiment 3 shows the results of a growth experiment using HeLa cells to verify that the hormone-dependent growth observed for the mammary cells really is specific for mammary cells and is not due to some unknown growth factor that was present which generally stimulates all types of cells. As can be seen from Fig. 3, the growth rate of HeLa cells was unaffected by addition of mammary-stimulating hormones, thereby verifying the specificity of the hormone stimulation for mammary cells.

Figure 4 shows the results of dilution experiments with fetal calf serum (FCS) and conditioned medium from a line of rat pituitary cells making rat prolactin and rat growth hormone (Yasumura *et al.*, 1966b). As previously mentioned, apparently fetal calf serum contains all the necessary hormones for mammary cell growth, since the cells will grow well in the basic synthetic medium with hypophysectomized dog serum if a small amount of fetal calf serum is added

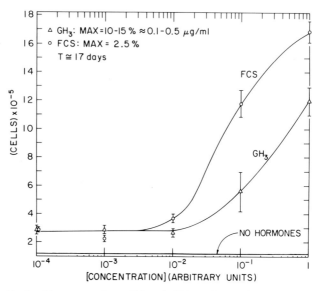

Fig. 4. Fetal calf serum and conditioned GH$_3$ medium dilution experiment done with uncloned mammary cells and Ham's F-10 culture medium with 10% hypophysectomized dog serum.

(2.5%) with *no* hormones (Fig. 2, positive control, highest horizontal line). Similarly, conditioned medium from the GH$_3$ rat pituitary cell line contains sufficient rat prolactin and rat growth hormone to stimulate growth of the mammary cells. Insulin was not used in this experiment, as conditioned GH$_3$ medium is capable of stimulating cell growth without the addition of insulin, although other experiments indicate that maximal stimulation by conditioned GH$_3$ medium requires the addition of insulin.

The scale used in Fig. 4 was arbitrarily fixed so that the maximum concentrations of fetal calf serum and conditioned GH$_3$ medium were identical to those used to grow up stock cultures of mammary cells and correspond to unity on the abscissa of the graph (2.5% fetal calf serum and 10–15% conditioned GH$_3$ medium). The concentrations were then varied downward from these maximum values in decade steps and the growth was measured. As in the case of the insulin dilution experiment, replicate flasks were plated with identical numbers of cells at the various concentrations of fetal calf serum and conditioned GH$_3$ medium. As can be seen from Fig. 4 the maximum concentrations of both FCS and conditioned GH$_3$ medium had to be diluted by about a factor of 100 to completely abolish the hormone stimulatory effect on the growth of the mammary cells. Complement fixation assays run on the conditioned GH_3 medium indicated that it contained on the order of 1 µg/ml of rat prolactin. Therefore the results of the dilution experiment indicate that prolactin concentrations of the order of 10–100 ng/ml are still effective in stimulating some growth of the cells, and that one must go down almost to the 1 ng/ml level to totally stop growth of the cells.

Work is currently in progress to study the effects of hormone stimulation on DNA, RNA, and protein synthesis and to determine the molecular mechanism of hormone cell interaction involved in the stimulation of cell growth. Also, we plan to look for differentiated cell function in terms of hormonal stimulation of milk protein synthesis.

SUMMARY

A systematic technique for the isolation of hormone-dependent cell strains from corresponding hormone-dependent animal tumors has been described, and its application to the isolation of a hormone-dependent, clonal line of rat mammary cells has been discussed in detail.

A second technique has also been described for inducing hormone-dependent tumors in animals, and both techniques are currently being employed to isolate hormone-dependent cell lines from different types of endocrine tissue.

The ultimate objective of these studies is to understand the mechanism of hormone stimulation of target endocrine tissue by pituitary tropic hormones.

The authors wish to thank Mr. John Nove and Miss Carol Roberts for their technical assistance, Professor D. F. Steiner, University of Chicago, for providing purified bovine insulin and proinsulin, and Dr. L. M. Sherwood, Beth Israel Hospital, Boston, Massachusetts, for supplying purified samples of human placental lactogen (HPL). The authors also wish to thank the National Institute of Metabolic Diseases and Arthritis and Dr. Albert Parlow, Harbor General Hospital, San Pedro, California, for supplying us with rat prolactin and rat growth hormone.

REFERENCES

AUGUSTI-TOCCO, G., and SATO, G. (1969). Establishment of functional clonal lines of neurons from mouse neuroblastoma. *Proc. Nat. Acad. Sci. U.S.* **64,** 311–315.

BENDA, P., LIGHTBODY, J., SATO, G., LEVINE, L., and SWEET, W. (1968). Differentiated rat glial cell strain in tissue culture. *Science* **161,** 370–371.

BUONASSISI, V., SATO, G., and COHEN, A. I. (1962). Hormone producing cultures of adrenal and pituitary tumor origin. *Proc. Nat. Acad. Sci. U.S.* **48,** 1184–1190.

Dent, J. N., Gadsden, E. L., and Furth, J. (1955). On the relation between thyroid depression and pituitary tumor induction in mice. *Cancer Res.* **15,** 70–75.

Dent, J. N., Gadsden, E. L., and Furth, J. (1956). Further studies on induction and growth of thyrotropic pituitary tumors in mice. *Cancer Res.* **16,** 171–174.

Furth, J., and Sobel, H. (1947). Neoplastic transformation of granulosa cells in grafts of normal ovaries into spleens of gonadectomized mice. *J. Nat. Cancer. Inst.* **8,** 7–16.

Furth, J., and Kim, U. (1960a). Relation of mammary tumors to mammotropes. I. Induction of mammary tumors in rats. *Proc. Soc. Exp. Biol. Med.* **103,** 640–642.

Furth, J., and Kim, U. (1960b). Relation of mammary tumors to mammotropes. II. Hormone responsiveness of 3-methylcholanthrene induced mammary carcinomas. *Proc. Soc. Exp. Biol. Med.* **103,** 643–645.

Furth, J., and Kim, U. (1960c). Relation of mammary tumors to mammotropes. IV. Development of highly hormone dependent mammary tumors. *Proc. Soc. Exp. Biol. Med.* **105,** 490–492.

Puck, T., and Kao, F. T. (1967). Genetics of somatic mammalian cells. V. Treatment with 5-BUDR and visible light for isolation of nutritionally deficient mutants. *Proc. Nat. Acad. Sci. U.S.* **58,** 1227–1234.

Rosenthal, M. D., Wishnow, R., and Sato, G. (1970). *In vitro* growth and differentiation of clonal populations of multipotential mouse cells derived from a transplantable teratocarcinoma. *J. Nat. Cancer Inst.* **44,** 1001–1014.

Shin, S. (1967). Studies on interstitial cells in tissue culture: Steroid biosynthesis in monolayers of mouse testicular interstitial cells. *Endocrinology* **81,** 440–448.

Shin, S., Yasumura, Y., and Sato, G. (1968). Studies on interstitial cells in tissue culture. II. Steroid biosynthesis by a clonal line of rat testicular intestinal cells. *Endocrinology* **82,** 614–616.

Tashjian, A. H., Jr., Yasumura, Y., Levine, L., Sato, G., and Parker, M. L. (1968). Establishment of clonal strains of rat pituitary tumor cells that secrete growth hormone. *Endocrinology* **82,** 342–352.

Turkington, R. W. (1968). Hormone induced synthesis of DNA by mammary gland *in vitro*. *Endocrinology* **82,** 540–546.

Yasumura, Y., and Sato, G. (1967). Prolonged cultivation of a pure strain of ACTH secreting pituitary tumor cells. *Abstr., 49th Meet. Endocrine Soc.*

Yasumura, Y., Buonassisi, V., and Sato, G. (1966a). Clonal analysis of differentiated function in animal cell culture research. I. Possible correlated maintenance of differentiated function and the diploid karyotype. *Cancer Res.* **25,** 529–535.

Yasumura, Y., Tashjian, A. H., Jr., and Sato, G. (1966b). Establishment of four functional, clonal strains of animal cells in culture. *Science* **154,** 1186–1189.

The Substratum for Bone Morphogenesis[1]

MARSHALL R. URIST

Bone Research Laboratory, UCLA School of Medicine, Los Angeles, California 90024

INTRODUCTION

Activated by injury, transplantation, or explantation, and provided with a substratum of bone matrix, mesenchymal cells of the musculoskeletal system express determination for differentiation of cartilage, bone, and bone marrow. Determination might conceivably stem from a substance which turns relevant genes on and off, or from inactive (masked) mRNAs, but whatever the mechanism the fact is that determination for differentiation of bone is held in abeyance in somatic mesenchyma throughout all the postfetal life. Permission to express this determination is granted to mammalian mesenchymal cells in: (a) intramuscular implants of a bone matrix *in vivo* (Urist, 1965); (b) fragments of muscle and bone matrix in Millipore chambers (Buring and Urist, 1967a); (c) explants of muscle upon a substratum of bone matrix in tissue culture (Urist and Nogami, 1970).

In transplants and explants of mesenchymal cells, the presence of a substratum first facilitates locomotion, then guides migration and proliferation, later permits aggregation and differentiation, and finally allows organ development. A grid made of plastic or glass and other nonbiologic materials, and fibrin clots or fish scales and similar biologic materials, will serve as a substratum for migration and proliferation of mesenchymal cells, but no other cell differentiation ensues (Weiss, 1961). Substrata prepared from extracted tendon collagen not only orient cell migration and proliferation, but also appear to condition the nutritional requirements of embryonic somite cells for differentiation of muscle (Königsberg and Hauschka, 1965). Substrata prepared from demineralized hard tissues not only orient new growth and permit differentiation of bone tissue, but also supports bone morphogenesis (Urist et al., 1968).

---

[1] This work was aided by a grant-in-aid from The John A. Hartford Foundation, Inc., and in part by grants-in-aid from the USPHS, National Institute of Dental Research (DE-02103), Ayerst Laboratories, Inc., and the Orthopedic Research and Education Foundation; and in part by a contract between the U. S. Army Research and Development Command (DA-49-193 MD-2556) and the University of California.

Under the guidance of bone morphogenetic substratum, new populations of mesenchymal cells are permitted to develop and at the same time are restricted to a specific line of specialization. In one respect the substratum performs the mechanical function of a railroad bed, which restricts movement of populations along a specified route. In another respect, the substratum is far more complex in physicochemical structure and far more influential than a mechanical road bed. In morphogenesis the substratum accelerates cell differentiation and promotes new growth like a chain reaction (Huggins, 1969) in which feedback from the developing whole integrates each subsequent step.

According to modern concepts of cell differentiation, substrata play no more than a permissive (supportive) role in morphogenesis. The genome plays the principal role; encoded in the genome is the nucleotide ordered sequence for determination of bone development. Once determination is established, a cell population develops osteogenetic competence (Urist et al., 1969b), the not-yet-activated state of readiness to respond to a bone morphogenetic substratum. Once osteogenetic competence is developed, there is no apparent target for a soluble diffusible inducer or agent of transmission of special instructions to the genome, and no reason to recall chemical theories of embryonic induction to explain bone morphogenesis.

In previous papers, a bone induction principle, defined as the product of interaction of extracellular products of mesenchymal cells with the chemical components of the substratum (Urist et al., 1967), was presented to explain bone morphogenesis. Nothing was found to suggest that a soluble diffusible component of bone matrix was responsible for bone morphogenesis, notwithstanding the occurrence of transfilter bone induction (Buring and Urist, 1967a). Consequently, the solid structure of the matrix was described as an "inductive substratum" in accordance with the terminology of Holtfreter (1968), who refers to solid substances as inductors and soluble substances as inducers. In the present communication, bone matrix is referred to as a "morphogenetic substratum" to emphasize the point that the solid structure supports morphogenesis, not merely histogenesis. A series of morphological and biochemical observations are presented below to postulate that *calcifiability* and *insolubility* are coupled properties of the organic matrix of bone and other hard tissues. In bone morphogenesis this property is as specific for performance of the permissive role of the substratum as the nucleotide sequence is for the performance of the principal role of the genome.

MATERIALS AND METHODS

Detailed accounts of methods of preparation of a bone morphogenetic matrix are described in previous articles (Buring and Urist, 1967a; Dubuc and Urist, 1967; Urist et al., 1968; van de Putte and Urist, 1966). Consistently reproducible results were obtained in rats with allogeneic intramuscular implants of bone matrix when: (1) the bone was demineralized in 0.6 N HCl immediately after excision (avoid freezing and thawing); (2) the temperature of the solution was 2°C continuously and stirred magnetically (EDTA, which denatures unfixed proteins must be avoided); (3) demineralization was accomplished within 3-4 days (longer periods of exposure to acid gelatinize the matrix). The demineralized cortical bone was washed in cold salt solution, implanted in the anterior abdominal wall, excised at 5-day intervals after the operation, and examined by histological, radiographic, histochemical, autoradiographic, and analytical chemical methods described in previous publications (Urist et al., 1967, 1968, 1969a,b, 1970; Yeomans and Urist, 1967).

Matrix from rachitic, lathyritic, and penicillaminated rat bone was prepared by the above-described method and implanted for comparison. The sources of the bone and experimental designs are recorded below in a running account of the results. The yield of new bone from both normal and abnormal sources of matrix were measured in terms of milligrams of ash per gram of preimplanted matrix at an arbitrary interval of 28 days after implantation. Before implantation, the matrix was ashless. After implantation, the weight of apatite mineral in new bone, in the incinerated implant, was dependent upon the quantity of new bone and directly proportional to the mass of preimplanted matrix (Urist et al., 1970).

For electron microscopic investigations, a series of implants of normal bone matrix were fixed supravitally by perfusion with 3.0% glutaraldehyde buffered with 0.33 M cacodylate, excised, and postfixed in 1.33% osmium tetroxide. Three unimplanted segments were similarly fixed and postfixed for examination. Both the unimplanted and implanted segments were cut into wedge-shaped fragments with the apex pointing toward the medullary space, dehydrated in solutions of alcohol of increasing concentrations, transferred to propylene oxide, impregnated with epoxy resin, and embedded in Epon. Polymerization was produced by heating at 60°C for 48 hours.

Thick sections were cut with glass knives and stained with toluidine blue to locate specific areas of interest with the light microscope. Thin sections were cut undecalcified from selected blocks trimmed down to less than 1.0^3 mm in volume. Two staining procedures were employed as follows: Grids were floated on saturated uranyl acetate in aqueous solution for 3.5 minutes followed by lead citrate for 3.5 minutes for impregnation of cell organelles and membranes, or exposed to 5% PTA for impregnation of collagen fibrils of old matrix and cement lines.

RESULTS

Cell Differentiation and Bone Morphogenesis in Implants of Bone Matrix

Eighty samples of demineralized normal adult rat cortical bone matrix were implanted in the anterior abdominal muscles of young rats and recovered at 5-day intervals between 0 and 40 days. Between 0 and 5 days, amoeboid mesenchymal cells (wandering histiocytes), migrated from the recipient muscle into the old marrow vascular channels of the old matrix, or formed an envelope of connective tissue around the implant. The amoeboid cells became fixed mesenchymal cells (FMC) and began to proliferate in the interior of the implant in the interval between the 5th and 10th days. Between the 10th and the 15th day, FMC fused to form multinucleated giant cells. These cells were 3 to 10 times larger than typical osteoclasts and were termed *matrix-clasts* because of their association with demineralized dead matrix, not with demineralized living bone. In 10-day implants, matrix-clasts had 50–500 nuclei and were large enough to fill the length of an old vascular channel or the volume of an entire excavation chamber measuring 350–3500 μ in diameter. In an excavation chamber, in a typical 10-day implant, the ratio of multinuclear to mononuclear cells was over 10:1.

In 15-day implants, at the time of the appearance of the first deposits of bone, large matrix-clasts were located mainly in deeper newer areas of cell invasion. In larger old excavation cavities, the matrix-clasts were progressively smaller in size, measuring 100–500 microns in diameter, and the ratio of multinuclear to mononuclear connective tissue cells was reversed to 1:10. The relatively small matrix-clast, and the low multinuclear–mononuclear cell ratio of 1:100, were generally associated with areas of new bone. In view of progressive reduction in numbers of nuclei immediately before the appearance of bone, absence of any evidence of degenerative change or pycnosis in the matrix-clast, and juxtaposition of mononuclear

CELL POPULATIONS IN THE INTERIOR OF IMPLANTS OF BONE MATRIX IN MUSCLE

Days postop.	Area of Matrix Resorbed	%	Total cell count per high power field, in 10 typical excavation chambers	Estimate of total no. of mitotic divisions of the average ^{3}H-thymidine labeled mesenchymal cell	Estimates of differential cell count, per 10 typical high power fields
5		2	5 ± 3	0	60 Polymorphonuclear leucocytes 15 < Small lymphocytes / Plasma cells 5 Red blood cell debris 20 Ameboid mesenchymal cells — 100
10		11	34 ± 8	1-2	25 Endothelial cells and RBC 25 < Plasma cells and / Ameboid mesenchymal cells 50 < Multinucleated giant cells / Proliferating mesenchymal cells
15		20	108 ± 12	2-3	45 Endothelial cells and RBC 50 < Multinucleated giant cells / Proliferating mesenchymal cells / Hypertrophied mesenchymal cells 5 < Chondroblasts and cytes / Osteoblasts and cytes
20		36	282 ± 36	3-4	60 Endothelial cells and RBC 30 < Multinucleated giant cells / Proliferating mesenchymal cells 10 < Chondroblasts and cytes / Osteoblasts and cytes / Osteoclasts
40		90	1040 ± 98	4-6	75 < Bone marrow / Endothelial cells, and RBC 10 < Multinucleated cells / Proliferating mesenchymal cells / Fibroblasts 15 < Chondroblasts and cytes / Osteoblasts and cytes

FIG. 1. Diagrammatic representation of correlated observations on changes in differential cell count, percent of the total volume of matrix resorbed by matrix-clasts, total cell counts, number of mitotic divisions of average thymidine-^{3}H labeled cell, at 5-day intervals over a period of 40 days post-implantation of demineralized bone matrix in muscle. Through these changes in cell populations, the implanted matrix is replaced by woven bone which in turn is remodeled to form a sphere-shaped lamellar bone ossicle filled with hematopoietic bone marrow.

osteoprogenitor cells in large excavation chambers, the latter were assumed to arise from nuclei pinched off from the cytoplasm of the matrix-clast.

No uptake of thymidine-^{3}H or evidence of mitotic division was noted in a matrix-clast injected at 10 days, but labeled nuclei were noted in matrix-clasts in implants injected with the nucleoside 96 hours previously, and in osteoprogenitor cells of implants injected

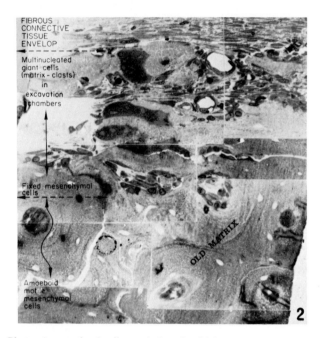

FIG. 2. Photomicrograph of cell populations in old bone matrix, 10 days after implantation in muscle, perfused with 3.0 percent glutaraldehyde in cacodylate buffer at pH 7.4. Note stratification of the different cell types in three depths of penetration of old matrix. Toluidine blue stain.

with the nucleoside at 15 days. Thus, the matrix-clasts which burrowed excavation chambers in the interior of the implant were in position in both time and space to form from fused mesenchymal cells, to carry collagenolytic enzymes to the substratum or matrix, and, to pinch off nuclei and produce osteoprogenitor cells, which in turn undergo mitotic division to produce the large population of cells engaged in osteogenesis. That perivascular mesenchymal cells which enter the area close behind the matrix-clast were a secondary source of osteoprogenitor cells, was not excluded, but circumstantial evidence pointed to the matrix-clast as a major source (Figs. 1 to 6).

Correlated observations on: (a) percent of the volume of collagenous matrix resorbed by matrix-clasts; (b) total cell population counts; (c) average number of mitotic divisions; and (d) differential cell counts summarized in Fig. 1, suggested that proliferation of new mesenchymal cells in the interior of the implant and resorption of old collagen by matrix-clasts were characteristic features of the preosseous phase of bone morphogenesis. Approximately 2% of the matrix was resorbed on the fifth day before, 11% by the 10th day after matrix-clasts form, and at least 20% by the 15th day at the time of the first appearance of osteoblasts and osteocytes in deposits of new bone. New bone did not form anywhere in areas of unresorbed matrix. Consequently, resorption of matrix appeared to be necessary to produce the space for the 20-fold rise in total cell population preceding osteogenesis. A larger than 200-fold increase in total cell population occurred after the 20th day, when the initial deposits of woven bone

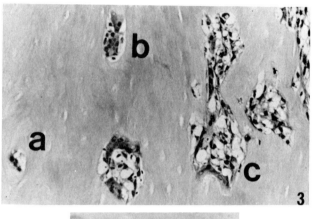

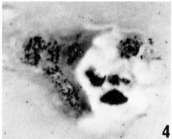

FIG. 3. Photomicrograph of bone matrix 10 days after implantation showing: *a*, fixed mesenchymal cells in old vascular channels; *b*, large matrix-clasts in small excavation chambers; *c*, small matrix-clasts in large excavation chambers.

FIG. 4. Autoradiograph of a large matrix-clast in a new excavation chamber at 15 days post-implantation, formed from mesenchymal cells in an animal treated with thymidine-^{3}H 96 hours previously; implants in animal treated 24 hours previously had labeling only in mesenchymal cells. The high grain counts in the matrix-clast at 96 hours after injection suggested that the multinucleated cell does not synthesize DNA or undergo mitotic division, and may form by fusion of previously existing, thymidine-^{3}H labeled mononuclear cells.

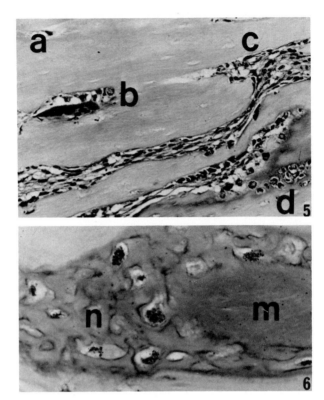

Fig. 5. Photomicrograph of bone matrix 15 days after implantation showing: a, fixed mesenchymal cells in old vascular channel; b, large matrix-clast in new excavation chamber; c, small matrix-clasts in larger excavation chambers; and, d, osteoblasts and new bone in the very largest of excavation chambers.

Fig. 6. Autoradiograph of new bone with thymidine-^{3}H labeled osteocytes in an implant of bone matrix, 20 days after the operation. The recipient was injected with thymidine-^{3}H, 4 days before excision. Animals injected with thymidine-^{3}H 24 hours before excision of the implant had no labeled osteoblasts or osteocytes; labeled bone cells (which do not synthesize DNA or undergo mitotic division) arise from thymidine-^{3}H labeled osteoprogenitor cells. Note: new bone with labeled osteocytes (n) surrounding old matrix (m).

were remodeled (resorbed and reconstructed), replaced with lamellar bone, and interspersed with a central pool of bone marrow.

Control implants of acid-alcohol demineralized bone matrix were resorbed and replaced by inflammatory round cells, giant cells, and fibroblasts. The end product was always a fibrous scar and never included cartilage or bone tissue.

Ultrastructure of the Substratum and Cell Interfaces before, during, and after Bone Morphogenesis

Electron micrographs of the implanted matrix revealed collagen fiber bundles densely packed in layers arranged alternately in transverse, oblique, and longitudinal directions. Old osteocyte lacunae were almost empty but contained variable quantities of amorphous cell debris. Old osteocyte canaliculi branching throughout the multilayered structure of the bone collagen, contained a granular electron dense material. Fragments of degraded cell membranes, adhering to the old matrix, were generally visible at 5 days but gradually disappeared at later intervals after implantation (Fig. 7).

First cell populations to react to the substratum. Electron micrographs of the amoeboid mesenchymal cells migrating along surfaces of old matrix at about 5 days after implantation demonstrated the morphology of the initial reaction to the substratum. These cells floated in finely granular material which covered all exposed surfaces of the implanted matrix. Their plasma membranes followed an undulating irregular course; mitochondria, endoplasmic reticulum (ER), and Golgi cysternae were undeveloped; small and large vacuoles were

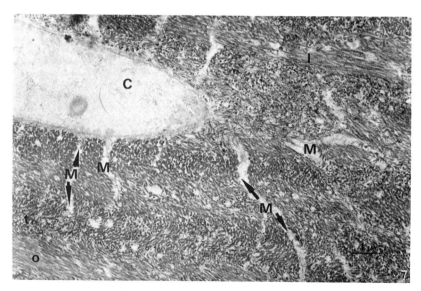

FIG. 7. Electron micrograph (*EM*) of the *interior* of the implanted matrix. c, Empty osteocyte lacuna; *M*, microcanaliculus; *t*, bone collagen fiber bundles in transverse; *o*, oblique; *l*, longitudinal section.

thinly distributed throughout the cytoplasm. The cytoplasm, indistinguishable from that of a macrophage, included electron-dense bodies and many vacuoles containing phagocytosed material (Fig. 8).

Subsequent cell reactions to the substratum. On the tenth day, daughter cells of migratory mesenchymal cells, covered all the exposed surfaces, outside and inside the implanted matrix. This cell population was mitotically active, densely packed, either spindle-shaped or polygonal, and called fixed mesenchymal cells (FMC). In FMC, mitochondria, dense bodies, glycogen bodies, and lyosomes were more prominent than in amoeboid cells, and no electron-dense material (ground substance) was interposed between surfaces of old bone matrix and the plasma membranes. In some places, the cell-substratum contact was less than 200 Å, and often closer than in any cell-to-cell contact of either loose, intermediate, tight, or specialized junctional complexes. In addition to this close order of contact, FMC inserted filopodial extensions into the opening of every available canaliculus of old matrix. The filopods were tubular extensions of plasmalemma, devoid of cytoplasmic organelles such as vacuoles, polysomes, mitochondria, but filled with a lacy network of microfilaments. Microtubules were not detectable within mesenchymal cell

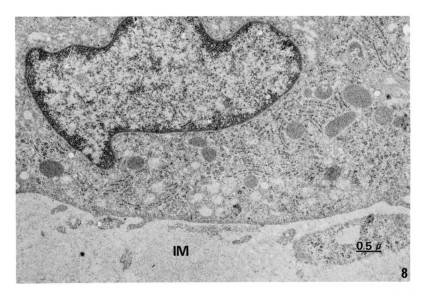

Fig. 8. Electron micrograph of migratory mesenchymal cell on the surface of old bone matrix (*IM*).

filopods. Thus, filipods were not ordinary mechanical protrusions of the cell into openings in the surface of the substratum, but retractable probes composed of structural elements of the cell membrane.

Unlike ordinary cell-to-cell tight junctions and junctional complexes which represent only limited areas of contact, cell-substratum areas of contact were extensive covering several μ^2 of cell and filopodial plasmalemma. In areas where the plasmalemma and substratum were sectioned in the long axis of collagen fibrils, contact appeared to be closer than 200 Å. There was no artifactual retraction of the plasma membrane from the substratum by the fixative solution as might be expected from a simple mechanical attachment. Attachments of FMC to the substratum of bone matrix, like selective affinity of homologous tissues, were repetitive or dynamic enough to make sheets of cells become polarized with respect to the structure of the substratum (Figs. 9–13). There was no deposit of ground substance and no new collagen between such sheets of polarized cells and surfaces of the matrix (as was noted in control implants of ceramic, silicon sponge, acid-alcohol inactivated matrix, and undemineralized bone).

Cell populations with ruffled borders in contact with the substratum of bone matrix. Interspersed with layers FMC in close contact with the substratum, were loose aggregates of FMC and multinucleated giant cells. Fusion of plasma membranes between FMC produced matrix-clasts which had all ultrastructural features of osteoclasts of growing bone except that, as noted previously, the size and number of nuclei in the average cell were significantly larger. The term matrix-clast was applied to these large cells which were polarized by a ruffled border or undulating membrane, on the side in contact with old mineral-free matrix. On the side opposite the surface of the matrix, the cell membrane was smooth and surrounded by a capillary tuft. Between the folds of the undulating membrane were strands of old bone matrix collagen undergoing digestion. The nuclei of the matrix-clast were congregated in the center of the cell. The cytoplasm was densely packed with small and large mitochondria. Between the collagen fibers and the surface of the undulating membrane was a PTA-staining granular nonfibrillar mucoid material measuring more than 500 Å in thickness. Thus, the cell-substrate contact between a matrix-clast and the bone collagen was characteristically a wavy fringe filled with partially digested collagen fibers (Fig. 14).

Condensation of ground substance between cartilage cells and the

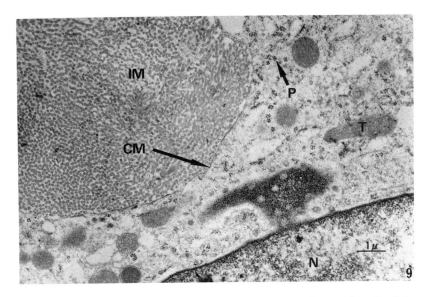

FIG. 9. Electron micrograph of bone matrix (*IM*), and fixed mesenchymal cell interface in cross section, at 10 days post implantation. *CM*, plasma membrane; *P*, polysome; *N*, nucleus; *T*, mitochondria.

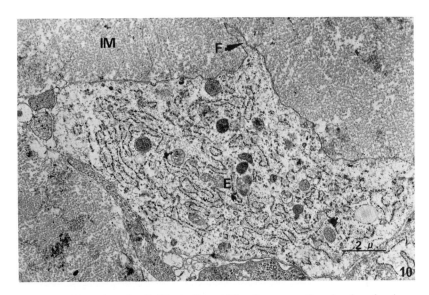

FIG. 10. Mesenchymal cell (*E*), in the middle of an excavation chamber showing a filopodial extension (*F*) completely filling a microcaniculus in old matrix (*IM*).

SUBSTRATUM FOR BONE MORPHOGENESIS 137

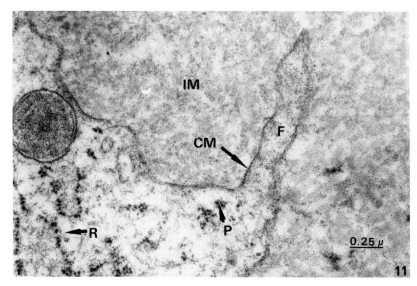

FIG. 11. Higher power magnification of the filopodial extension, F, shown in Fig. 10. Polysomes (P), and other organelles do not flow into the filopod. IM, Old bone matrix collagen fibrils in cross section; CM, cell membrane; R, ergastoplasm.

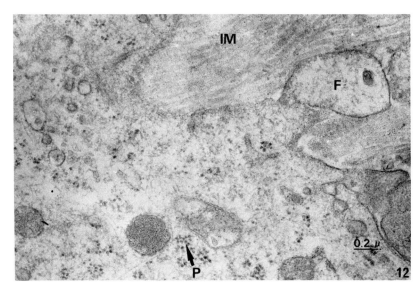

FIG. 12. Higher power electron micrograph of the interior of a filopod (F) in cross section to show absence of polysomes and presence of organic material with the electron density and structure of plasma membrane. IM, Old collagen; P, polysomes.

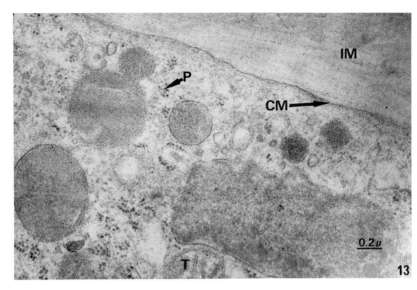

FIG. 13. Electron micrograph of the substratum-mesenchymal cell interface in higher power magnification to show old bone collagen (*IM*) in longitudinal section. *CM*, Line of tight contact with plasma membrane; *T*, mitochondrion; *P*, polysome.

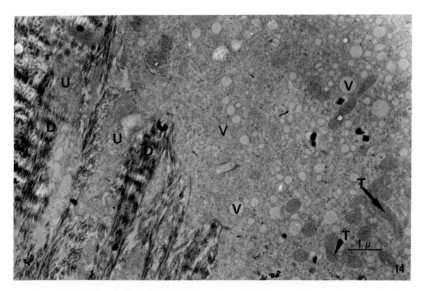

FIG. 14. Electron micrograph of folds of undulating membrane of a matrix-clast (*U*). *D*, Bone collagen fiber bundles undergoing digestion; *V*, vacuoles; *T*, mitochondria.

substratum of bone matrix. Beginning as early as 10 days after the operation, FMC proliferating in deep recesses of the implanted matrix unpenetrated by capillary sprouts did not lose intercellular plasma membranes, or fuse to form multinuclear cells. Instead, aggregates of FMC developed plasma membranes with an electron-dense cortex and capsule, well-developed rough endoplasmic reticulum, a prominent Golgi apparatus and deposits of glycogen. These cells, which had all the ultrastructural features of typical chondrocytes became cemented to the substratum of old matrix by a layer of condensed ground substance measuring about 2000 Å in thickness. No part of the plasma membrane cortex of a cartilage cell was closer than 0.5–1.0 μ from the implanted matrix (Fig. 15).

Points of contact between cells diminished as chondrocytes became spherical in shape and separated farther apart by secretion of increasing quantities of cartilage matrix. As the matrix between the cells increased in quantity, chondrocytes became more round and the nucleus moved to a more central location. An enlarged Golgi apparatus appeared in close juxtaposition to the nucleus while the cytoplasm lost glycogen. The new matrix consisted of widely dispersed

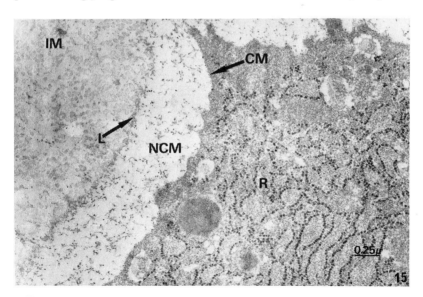

Fig. 15. Electron micrograph of cell substratum interface between old bone matrix and a young chondrocyte. *IM,* Old collagen fibers in cross section; *L,* cement line; *NCM,* new cartilage matrix; *CM,* cortex of plasma membrane of a chondrocyte; *R,* ergastoplasm.

thin filaments and fibrils of collagen measuring 50–500 A in diameter which were suspended in a watery bed of amorphous material stainable as chondromucoprotein. The above ultrastructural characteristics clearly distinguished the chondroblasts from both fibroblasts and osteoprogenitor cells.

Condensation of ground substance between bone cells and the implanted substratum. At 15 days after the operation, in areas of the implant, previously excavated by matrix-clasts and sprouting capillaries, osteoblasts, and new bone tissue began to differentiate. The ultrastructure of the bone cell in the newly woven bone in an implant of bone matrix was the same as in the metaphysis of a normal growing bone; osteoblasts had rough ER concentrated at one side, generally the side adjacent to calcifying new collagenous bone matrix; Golgi apparatus was prominent and occupied a juxtanuclear position; extracellular collagen fibrils were 2 to 3 times more densely packed than those secreted by young fibroblasts (Figs. 16 and 17).

The new mineral consisted of needles of apatite clustered within and around osmiophilic microglobules which appeared to condense at randomly scattered points between bundles of new collagen fibrils, but always several microns from the surface of the cell membrane. Lacking microsomal bodies, the microglobules of organic material in which the earliest deposits of apatite microcrystals appeared, were not simple fragments of cytoplasm but had the same structure as microglobules in calcifying epiphyseal cartilage. The volume, number, and distribution were the same as that of elastic fibers in noncalcifying fibrous connective tissue. The *earliest* deposits of mineral identified by EM fixed in solutions of neutral pH were oriented, not to cross-banded collagen fibrils, but to microglobules of electron-dense material which would have had to crystallize like elastic fibers in the ground substance between bundles of collagen some distance from surfaces of cells. After clusters of apatite appeared in microglobules of organic material, the surrounding collagen fibers began to calcify. Calcification spread from a constellation of calcified microglobules to intervening bundles of new bone collagen, but always excluded the thick layer of loosely woven collagen immediately adjacent to the plasma membranes. This layer of noncalcifying collagen and ground substance persisted around osteoblasts and osteocytes of the deposits of new bone at all stages of development (Fig. 18).

Crystal formation, first in microglobules and later in spaces between bone collagen fibrils, but always at a distance from plasma

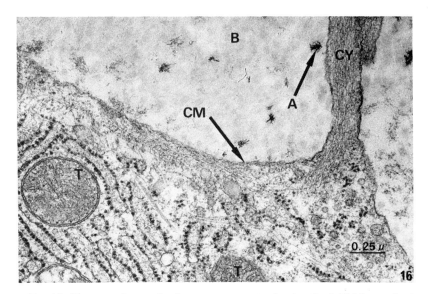

Fig. 16. Electron micrograph of a young osteocyte showing the initial locus of calcification in widely scattered electron-dense globules of osmiophilic material. R, Ergastoplasm; T, mitochondria; CM, cell membrane; B, new bone collagen; A, apatite microcrystallites; CY, cytoplasmic process.

membranes, suggested that the mineralization process was controlled by the osteoblast at an earlier stage beginning with processes of the intracellular synthesis and secretion of calcifiable matrix. Individual microcrystallites of apatite were too small to be resolved by EM even at magnifications of 100,000×. The overlapping of structural images of the crisscross lamellae of bone collagen, in section, as thin as 1.0 μ was too great to localize the mineral anywhere except *between* and *in* the surfaces (not in the interior) of a collagen fibril. No vestige of old mineral remained and no remineralization occurred within the substance of the old bone collagen. Only new bone matrix mineralized and the new bone deposits became attached to the wholly demineralized old matrix by an electron-lucent amorphous layer of cement substance measuring more than 1 μ in thickness (Fig. 19).

Points of Loose Contact between Osteoprogenitor Cells and Old Bone Matrix

In electron micrographs, osteoprogenitor cells appear in large excavation chambers in juxtaposition to matrix-clasts and osteoblasts, in

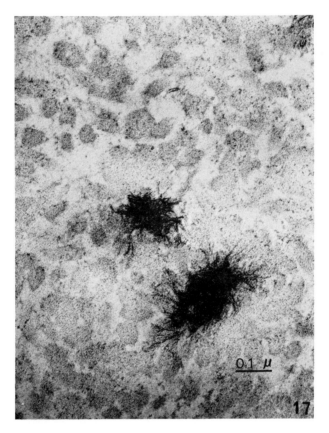

Fig. 17. Higher power electron micrograph of the initial locus of calcification showing apatite microcrystallites in globules of osmiophilic material.

areas between capillary loops and new bone. Geographically, these areas correspond to areas occupied by thymidine-^{3}H-labeled connective tissue cells in the interstices of deposits of new bone. Other cells, matrix-clasts and osteoblasts, and osteocytes (which did not undergo mitotic division or utilize thymidine-^{3}H) had a characteristic ultrastructure and were easily differentiated from osteoprogenitor cells. Endothelial cells and perivascular connective tissue cells of bone tissues, utilized thymidine-^{3}H, but were smaller in size and not in apposition to old bone matrix. Ultrastructurally, the cytoplasm of the typical osteoprogenitor cells was almost as rich in mitochondria and lyosomes as an osteoclast and in addition had some of the

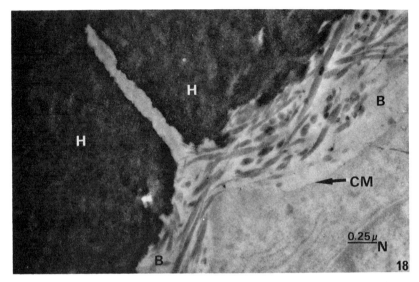

FIG. 18. Electron micrograph of interfaces between an osteocyte cell membrane (CM), moat of uncalcified new collagen (B), and heavily calcified new bone collagen (H). M, Microcanaliculus of young osteocyte; N, nucleus.

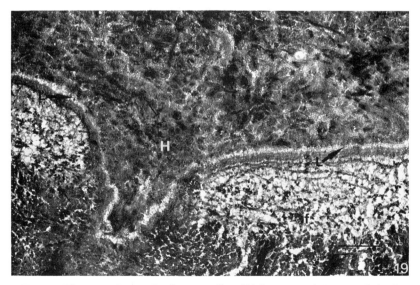

FIG. 19. Electron micrograph of cement line (L) between substratum of the implanted old (f) and deposit of heavily calcified new (H) bone matrix. Glutaraldehyde and osmic acid fixation. Phosphotungstic acid stain.

coarse ER of the osteoblast. A film of granular material, presumably ground substance, was interposed between small points of loose contact between plasma membranes and old bone matrix.

Sequence of Cell-Substratum Interfaces with Cell Differentiation

Figure 20 is a diagrammatic representation of the types and dimensions of interfaces developing before, during, and after differentiation of various specialized cells in bone morphogenesis. At 5 days, a granular film or ground substance filled the space between plasmalemma of motile mesenchymal cells; at 10 days, plasmalemma of FMC including filopodial extensions were in tight contact with the

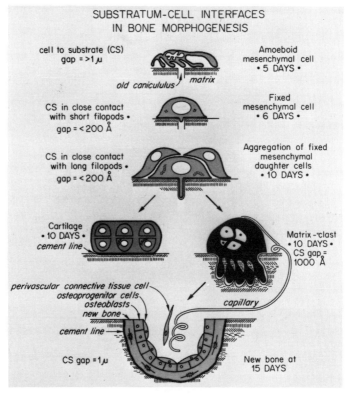

FIG. 20. Diagrammatic illustration of structure and dimensions of interfaces between cells and substratum in successive phases of bone morphogenesis: distant contact at 5 days; close contact from the 6th to 10th day; intermittent contact of undulating membrane of matrix-clasts from the 10th to the 15th day; cement lines between cartilage and matrix, or new bone and old matrix after 15 days.

TABLE 1
Yields of New Bone from Cold 0.6 N HCl-Demineralized Normal and Abnormal Matrix

Preparation	Chemical morphology	Yield of new bone (mg bone ash per gm of preimplanted matrix)
Normal cortical bone	Demineralized cross-linked collagenous matrix	450 ± 50
Normal dentin	Demineralized cross-linked collagenous matrix	500 ± 50
Pencillaminated cortical bone	Thiazolodinized cross-linkages	72 ± 14
Lathyritic cortical bone	Defective cross-links between ϵ-amino and aldehyde groups	60 ± 10
Vitamin D and phosphorus deficiency rachitic osteoid and cortical bone	Uncalcified cross-linked collagenous matrix	80 ± 20

substratum with no demonstrable intervening material, while the undulating membranes of matrix-clasts were filled with wavy fringe of partially digested collagen fibers; at 15 days, osteoprogenitor cells with points of loose contact between plasmalemma and old matrix appeared in areas adjacent to deposits of new bone; at 20 days, a layer of condensed ground substance was deposited by cartilage cells, or cement substance was secreted by bone cells which sealed all of the surface area between old and new matrix. Viewed in the light of the Weiss (1961) concept of contact guidance, all of the above-described cell contacts, except the last one, were assumed to be adhesive interfaces but always of a transitory nature.

Yield of New Bone from Normal Matrix

Table 1 summarizes the results of experiments with over 200 intramuscular implants of preparations of bone and dentin matrix which produce a high yield of new bone within 4 weeks in allogeneic rats. Negative results occurred in less than 0.5% and these were attributable to infection or ischemic necrosis of muscle. The yield of new bone from demineralized bone matrix was 450 ± 50 mg of new bone ash per gram of matrix (Fig. 21). Slightly higher yields of new bone 500 ± 50 were obtained from implants of dentin matrix.

Dentin matrix is more permeable to cold HCl and more firmly cross-linked even than compact bone matrix. In both bone and dentin, the collagen was relatively insoluble in 5 M guanidine, and therefore con-

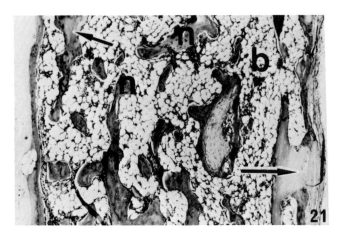

Fig. 21. Photomicrograph of the implant of 0.6 N HCl demineralized normal matrix shown 6 weeks after the operation. Remnant of demineralized normal old acellular matrix is shown by arrows. n, Calcified new bone matrix; b, new bone marrow.

sidered more densely cross-linked than tendon or skin collagens (Urist et al., 1968, 1969a), but, as will be discussed further, components of the physicochemical structure, other than the cross-links, comprise the bone morphogenetic substratum.

Preparations of Cold HCl-Demineralized Matrix which do not Impose Bone Morphogenetic Pattern upon Mesenchyme

Vitamin D and phosphorus-deficiency rickets. Six samples of matrix prepared from tibial and femoral cortical bone of rats with high calcium, low phosphorus vitamin D-deficiency rickets, were prepared by decalcification in 0.1 N HCl at 2°C for 8 hours with stirring. Thirty cylinders of demineralized rachitic bone were implanted in the anterior abdominal wall of normal allogeneic rats for 6 weeks. The rachitic rats were supplied by Dr. Hector DeLuca and shipped in the frozen state from the vitamin D-deficient rat colony of the University of Wisconsin. The long bones were severely rachitic, yet just as insoluble in 5 M guanidine as normal matrix.

The preimplanted bone matrix consisted of a small central area of demineralized normal bone (developed in the prerachitic period of life), and a large area of uncalcified osteoid (developed in the postrachitic period of life). Six weeks after implantation of this material, the central area of demineralized normal matrix had produced a small deposit of new bone and bone marrow. The peripheral area of osteoid tissue did not mineralize but was rapidly resorbed and replaced by fibrous tissue (Figs. 22–24). The yield of new bone from the total sample of rachitic and nonrachitic bone was only 7% of normal bone matrix. These observations suggested that only matrix that had been calcified beforehand, irrespective of the high strength of the cross-links of the collagen, had the bone morphogenetic property.

Demineralized penicillaminated bone matrix. Penicillaminated bone was obtained from 7-week-old rats reared on a diet containing 2% penicillamine in ground chow for a period of 4 weeks. The fully mineralized penicillaminated long bones were demineralized in 0.6 N HCl at 2°C with stirring for 24 hours. The bones were washed in normal saline, lyophilized, cut into diaphyseal cylinders 1.0 cm in length,

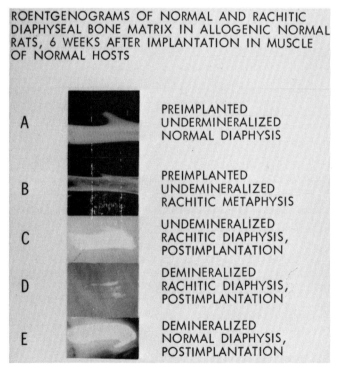

FIG. 22. Roentgenograms of preimplanted normal (A) and rachitic (B) tibial diaphyses including the attachment of the fibula. The area of deposits of new bone implants of rachitic (D) and normal (E) demineralized bone matrix are compared with undemineralized rachitic matrix (C) to show lack of either resorption or remodeling in the latter. Note the accurate reproduction of the shape of the fibular spur of normal matrix of implant (E), but not of the osteoid matrix of implant (D).

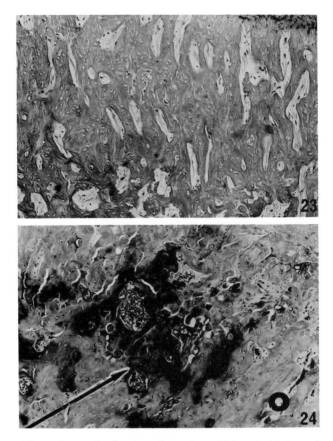

Fig. 23. Photomicrograph showing allogeneic rachitic osteoid free of any bone tissue, 6 weeks after implantation in the anterior abdominal wall of an allogeneic nonrachitic adult rat. Note the fibrous tissue in old vascular channels and acellular osteoid matrix. Hematoxylin-eosin and azure stain.

Fig. 24. Photomicrograph showing a deposit of new bone in an island of demineralized old matrix in the center of the implant of rachitic diaphyseal osteoid shown in Fig. 23. Bone morphogenesis is thus limited strictly to the area of remnants of normal nonrachitic matrix (arrow). No new bone formation occurs anywhere in the surrounding rachitic osteoid matrix. Note the osteocytes in new bone and the absence of osteocytes in lacunae of old osteoid matrix (O).

weighed, and stored. These preparations of bone matrix were tested in 5 M guanidine to ascertain the degree of penicillamination and found to be almost 8 × more soluble than normal control bone matrix (1 gm of matrix yielded 17 mg of hydroxyproline compared with the normal control matrix which yielded 2.4 mg).

Five penicillaminated diaphyseal bone cylinders of known weight were implanted in the anterior abdominal wall of 6 adult rats for a period of 42 days. One segment of 0.6 N HCl demineralized bone matrix prepared from long bones of normal 7-week-old animals was implanted in each recipient rat for a control. The implants were excised and examined by radiographic, histological, and ashing techniques to measure the yield of new bone per unit mass or preimplanted matrix. Table 1 summarizes the results of implantation of 30 samples of penicillaminated bone. After 40 days, the yield of new bone was 72 mg of new bone ash per gram of preimplanted matrix, only 16% of the yield of normal bone matrix. Histologically, there was no new bone in the area of the penicillaminated bone. Such scanty bone deposits as there were, were deposited in apposition to remnants of normal bone developed during the pre-penicillamine period of life.

Demineralized lathyritic bone matrix. Lathyritic bone matrix was obtained from 6-week-old rats, reared on a diet containing 5 gm of β-aminopropionitrile (hydrochloride) per kilogram of laboratory chow, for a period of 4 weeks. The periosteum and marrow-free diaphyseal cortical bone were rinsed in cold distilled water and demineralized in 0.6 N HCl at 2°C for only 2 days. The acid was removed by washing with water and the tissue was lyophilized and later implanted into the rectus abdominal muscles of allogeneic rats.

Before implantation the dimineralized lathyritic cortical bone contained inclusions of small islands of normal bone remaining from the preweaned period of growth. Following implantation, new bone appeared in the interior of small inclusions of normal bone matrix, while the large surrounding mass of lathyritic bone produced no new bone. As noted in Table 1, the yield of new bone measured in milligrams of ash per gram of preimplanted matrix was only about 13% of normal matrix. This small quantity of new bone ash was located strictly within islands of normal bone matrix. Lathyritic bone (from which large quantities of collagen and mucopolysaccharides are extractable in guanidine HCl), had no bone morphogenetic activity, either before or after exposure to these extractant solutions (Figs. 25 and 26).

DISCUSSION

The foregoing light microscopic, electron micrographic, and chemical observations suggest that the 3-dimensional complex structure of the substratum and no known single chemical component of bone

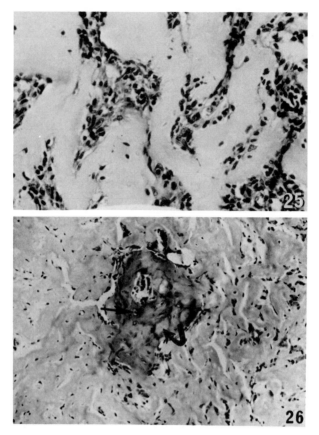

FIG. 25. Photomicrograph of demineralized lathyritic bone matrix free of any normal bone matrix, in the anterior abdominal wall of an allogeneic normal rat, 6 weeks after implantation. Only fibrous tissue histogenesis, and no bone morphogenesis, occurs in lathyritic bone matrix.

FIG. 26. Photomicrograph showing a deposit of new bone (straight arrow), upon an island of demineralized old normal matrix (curved arrow), surrounded by demineralized old lathyritic matrix, 6 weeks after implantation in the anterior abdominal wall of a normal adult rat. Bone morphogenesis does not occur anywhere in the lathyritic area of the implant.

tissue imposes a bone morphogenetic pattern on proliferating mesenchymal cells.

Kinetics of Cell-Substratum Interactions

The explosive increase in total cell population, the changing character of the differential cell counts, and the percentage of the volume of bone collagen resorbed before bone cells differentiate, point to a causal (albeit indirect) relationship between the implanted matrix and bone morphogenesis. Within 5 days after contact with the substratum, a cell population which carries the genome for synthesis of a wide variety of *noncalcifiable* extracellular substances, restricts its biosynthetic activities to production of three end products: uncalcified chondromucoproteins of cartilage, *calcifiable* extracellular substances of bone, and blood-forming bone marrow.

The number of cells in the *initial* population to interact with the substratum is relatively small, approximately 5 ± 3 per high power field in a typical section (on the 5th day) but a burst of mitotic activity (between the 6th and the 10th day) increases the number to 34 ± 8. The increase in mitogenic activity occurs concomitant with an increase in the rate of synthesis and a 6- to 10-fold rise in quantity of alkaline phosphatase activity (Huggins and Urist, 1970).

Inasmuch as no comparable elevation in alkaline phosphatase activity occurs in acid-alcohol inactivated bone implants, and insofar as close contact between migratory mesenchymal cells and old bone matrix develops only in implants of undenatured matrix, it is reasonable to assume that a bone morphogenetic pattern is transferred from the substratum to a relatively small number of fixed mesenchymal cells at about the 6th day. How the morphogenetic pattern is imprinted upon a small cell population and reprinted upon all of the larger cell populations developing after 6 days is a mystery. The facts in the case imply that a relay system exists and replication of intrinsic determinant with extrinsic substrata occurs in bone development in higher animals as in organ development in lower animals described by Harris (1968) and Trinkaus (1969). But nothing is known about replication of an extrinsic substratum such as the bone morphogenetic pattern or how it could be imprinted and reprinted in the extracellular matrix of several cell generations. When the bone morphogenetic imprint is established in the extracellular substance of a new cell population, reprints would be fed back from the developing whole at each subsequent phase, as shown in Fig. 27, to produce an ossicle complete with a medullary cavity and bone marrow.

Ultrastructure of the Bone Morphogenetic Interface

Electron micrographs of a bone implant disclose lines of contact between the bone matrix substratum and cell membranes as close or closer than the cell surface-to-surface contacts observed in embryonic connective tissues by Trelstad *et al.* (1966). These cell contacts in-

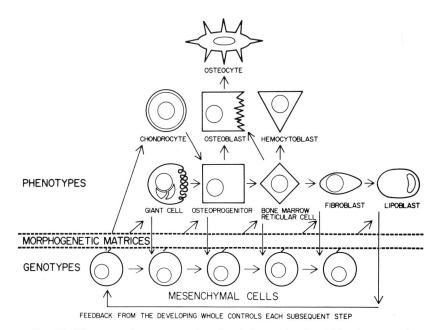

FIG. 27. Diagrammatic representation of a chain reaction in which a bone morphogenetic pattern is fed back from the developing whole to each subsequent step in bone development.

volve filopodial extensions containing microfilaments which increase the cell surfaces to cover large areas of the substratum. The combination of filopods and close contact polarizes the external and internal structure of the cell in relation to the substratum, but there is no dense material comparable to the structure of a zonula occludens at cell substratum interfaces. In fact, the initial contact between cells and old bone matrix is clearly dynamic that is, rapidly changing yet extremely close and therefore physicochemical in nature.

As noted previously, the earliest chemically detectable response of the cell population to contact with the substratum, is a rise in the level of alkaline phosphatase, and when measured in terms per unit weight of DNA in the system (Firschein and Urist, unpublished data), the elevation is absolute and not relative to increase in the cell population. Moreover, the rise continues throughout the interval between the 5th and the 20th days after implantation. In histochemical EM studies, Gothlin (personal communications) and Doty (personal communications), demonstrate alkaline phosphatase localized in the plasma membranes of osteoprogenitor cells, cartilage cells, and osteoblasts, but in none of the adjacent mesenchymal cells of osteoclasts. Whether alkaline phosphatase activity is associated with high energy phosphatase metabolism in general, or with collagen synthesis, or with calcification of bone matrix, or some other cell function, is not clear.

Granted that the substratum simply accelerates the existing enzyme biosynthetic machinery and does not induce construction of new machinery, we are now faced with the task of determining the nature and location of the accelerator. Some of the possible accelerators of enzyme synthesis, discussed in the literature on cell differentiation are: change in plasma membrane permeability (Fell, 1969); synchronous locking of cellular rhythms with terrestrial patterns (Goodwin, 1969); small quantities of a specialized product monitored by developing tissues (Zwilling, 1968); electrical currents including PN and NPN junctions (Becker and Murray, 1967); macromolecular complexing at inductive interfaces (Grobstein, 1967); ribonucleoproteins (Yamada, 1961); corticogenes (Nanney, 1968); enrichment of the nutrients in the environment (Lash, 1968; Ellison *et al.*, 1969). The resolving power of the best electron microscopes and application of improved autoradiographic and histochemical techniques of the future will be necessary to obtain more tangible information about the bone morphogenetic pattern.

Chemical Morphology of the Bone Morphogenetic Matrix

The matrix of bone contains fibrous proteins, glycoproteins, lipoproteins, salts, water, and various constituents of all connective tissues. The principal tissue-specific feature of the bone matrix is a particular proportion and arrangement of these constitutents. Dentin matrix, which has the same constituents as bone matrix, and in nearly the same proportions, has a different structural arrangement. In dentin the collagen fibers are as densely packed as in bone, but more firmly cross-linked and arranged around microtubules in a pattern that is highly permeable to solvents. After demineralization and implantation in muscle, both dentin and bone matrix produce new bone with equal consistency, but the yield is slightly higher with the former than the latter.

Intramuscular implants of bone matrix, prepared from rachitic, lathyritic, and penicillaminated bones demineralized under similar conditions, and compared with normal bone, present interesting information about the biologic properties of the bone collagen, and the difficulty of deciphering the bone morphogenetic pattern. Implants of uncalcified rachitic osteoid do not recalcify, and fail to produce new bone in muscle. Even osteoid tissue, not exposed to the demineralizing solution *in vitro*, fails to mineralize or to evocate osteogenesis. That which is present in normal bone matrix but lacking in osteoid and responsible for bone morphogenetic pattern, defies detection by ordinary chemical methods. Osteoid tissue, although uncalcified, has intra- and intermolecular cross-linkages, as firm as those of normal bone matrix. Osteoid tissue yields only 1.4 mg hydroxyproline per gram of matrix, about the same quantity as normal matrix (2.4 mg per gram of wet weight) after extraction with 5 M guanidine (author, unpublished observations). Osteoid tissue is not mineralizable until it becomes biochemically and physically altered; it fails to mineralize either *in vitro* or *in vivo* under conditions in which hypertrophic cartilage matrix always mineralizes. No obvious explanation for this fact is apparent, except that nonmineralizable osteoid (unlike demineralized bone matrix), is metachromatic, less polymerized, and may have an excess of reduced anionic radicals (Thomas, 1961). Thus, although the low solubility of the collagen in uncalcified osteoid and normal bone matrix is similar, the histochemical properties and physical configuration are sufficiently different to account for the lack of the morphogenetic pattern for normal bone development.

Matrix prepared from lathyritic and penicillaminated bone differs strikingly from normal bone matrix in solubility of the demineralized collagen in 5 M guanidine. Lathyritic matrix produces 7.0 mg (Strates and Urist, 1969), and penicillaminated matrix (unpublished data) produces 17.2 mg of hydroxyproline per gram of wet weight of tissue, about 3- and 8-fold, respectively, more than normal bone. Lathyrism and penicillamism affect much more than the solubility of the protein, but the emphasis in recent research is almost entirely on the changes in the properties of collagen. Penicillamine forms a thiazolidine ring with aldehydes, thereby blocking the process of cross-linking of collagen (Nimni, 1968). Lathyrogens produce defective or blocked cross-linkages in bone matrix possibly by inhibiting amino acid oxidase metabolism (Martin *et al.*, 1961).

Some components of the bone matrix are less important than others in the structure of the bone morphogenetic pattern, but the collagen is clearly essential. Collagen (which represents approximately 90% of the total dry weight of the organic matrix of cortical bone) must be carefully preserved to retain the bone morphogenetic property. Bone matrix collagen is more insoluble than collagens of noncalcifiable connective tissues in weak acids in the cold; recent work by Lane (1970), suggests that there are also differences in the arrangement of the sequences of certain amino acid constituents. Provided the native structure of calcifiable collagen including the stabilizing side chains are not degraded in the process, nearly every other component can be extracted without reducing the yield of new bone.

The solid state of the collagen, upon which bone morphogenesis depends, is sustained by intra- and intermolecular cross-links with strong covalent and numerous noncovalent bonds. The cross-links are so firm that weakly acidic solutions may denature or alter the physicochemical structure without increasing the solubility of the bone collagen. However, long exposure to concentrated solutions of hydrogen ions reduce the charge on carboxyl groups, break hydrogen bonds, destroy electrostatic linkages, and reduce van der Waals forces responsible for the stability of the collagen structure. Acid-alcohol solutions may neutralize the charge on amino groups and reduce the metal ion chelating or stabilizing action of the histidine residues of the bone collagen. The combination of acidic and alkaline solutions in sequence destroys all forces responsible first for the quarternary and then the tertiary (coiled-coil), and eventually the secondary structure of collagen.

Calcified tendons, cartilages, and arterial walls produce bone after implantation in muscle (Urist et al., 1968, 1969a) but whether the collagen or elastica of the *calcified* soft tissues are more insoluble than these proteins in *uncalcified* tissues is not known. The supernumerary cross-linkages of the collagen of calcified hard tissues may form as the tissues become calcifiable. At the present time, the structure of the calcification initiator, including its relationship to the cross-links in collagen is difficult to define. Information obtained from experiments on calcification in implants of cartilage, tendon, and aorta suggests that elastoid, and not any of its component parts, constitutes the initial locus of calcification (Urist and Adams, 1967; Bonucci, 1967). Elastoid is a mixture of calcium complexes of protein, lipoprotein, glycoprotein salts, and water.

In the cold, dilute HCl does not reduce morphogenetic activity of implants of bone matrix while the same solutions at 25°C or more, or stronger solutions of hydrogen ion, rapidly destroy the bone morphogenetic pattern. Other chemical agents that either block, alter, or denature the nitrogenous constituents of the bone collagen also destroy the bone matrix collagen as a substratum for new bone formation. Acidic solutions at room temperature which denature and transform the bone collagen into gelatin, nitric and nitrous acids which deaminate collagen; fluorodinitrobenzene (FDNB) which both denatures and dinitrophenolates the ϵ-amino groups of lysine and hydroxylysine; dilute alkali which neutralizes the charge on amino groups or ruptures cross-linkages; doses of ionizing radiation in excess of two million roentgen; and enzymes such as collagenases and pronase which digest the collagen molecule, completely destroy the morphogenetic property (Buring and Urist, 1967b; Urist et al., 1968). As more is learned about substances bound to the collagen of calcifying tissues, it might be possible to determine whether the fibrous protein or the components attached to it, or both, are the essential part of the bone morphogenetic pattern. Bone matrix contains as much as 2% acid insoluble protein other than collagen, but very little is known about the physiologic significance of this fraction.

Approximately 5% of the dry weight of bone matrix consists of glycoprotein. The bone collagen polymerizes or crystallizes in a ground substance consisting of glycoprotein, salts, and water. Once crystallized it is virtually impossible to separate the glycoprotein from the fibrous protein without use of extracting solutions which denature the bone collagen. Two of these solutions, EDTA and EGTA, diminish or destroy the bone morphogenetic pattern. Neutral salt solutions of calcium and other divalent metal ions form metal ion complexes with carboxyl groups of the bone collagen, extract a part of the glycoprotein of bone matrix, but do not destroy the bone morphogenetic pattern. Enzymes such as hyaluronidase, papain, and chymopapain which digest some of the bone glycoprotein without denaturing the total bone matrix do not destroy the bone morphogenetic pattern. These observations are interpreted as evidence that the glycoprotein is not the principal component of the morphogenetic pattern for bone formation. Bone sialoprotein represents 0.5% dry weight of the total glycoprotein and is extractable with 0.6 M EDTA or by proteolytic digestion and periodic oxidation, but this procedure denatures the bone collagen and destroys the bone morphogenetic property. Thus neither bone morphogenesis nor calcifiability can be attributed to the presence or absence of the sialoprotein component in an implant of bone matrix.

The dry weight of bone matrix contains less than 0.5% lipid. As noted in Table 2, extraction of lipid from demineralized matrix with ether, alcohol, acetone, chloroform-ethanol, hexachlorophene, dodecyl sulfate, and Tween 80, does not destroy the bone morphogenetic pattern. Destruction does occur when demineralization and extraction of lipid and lipoproteins are concurrent in acidic-solutions of alcohol, ether, or acetone; presumably an acid solution of a fat solvent ruptures both hydrophobic bonds and cross-links between molecules of collagen, and thereby destroys the bone morphogenetic pattern (Urist and Strates, 1970).

Transformation of soft tissues into a calcifying matrix occurs following uptake of large quantities of calcium by tendon and elastica (Urist and Adams, 1967). The binding of calcium to lipoproteins, glycoproteins and protein-proteins alters the arrangements of salts and water, and produces a complex material called elastoid. This complex is not elastin, a specialized protein which contains large quantities of valine, but it is like elastin insofar as it fails to diffract X-rays or produce a characteristic pattern. The calcium content of elastoid is 4–6 times higher than that of either elastica or collagen (Urist and Adams, 1967). To investigate the possibility that the calcium in elastoid may be the essential component of the bone morphogenetic pattern, rat bone matrix was demineralized in 0.6 M citric acid in the cold for periods as long as 30 days or until all except 0.003 mmole of calcium per kilogram of dry weight of the tissue was removed. This preparation of calcium-depleted bone matrix did impose the morphogenetic pattern on mesenchymal cells almost as well as the organic matrix with its full complement of protein-bound calcium (Urist et al., 1968).

After demineralization in cold dilute HCl and fixation in liquid nitrogen for freeze drying, the osteocyte lacunae are filled with variable quantities of cell debris. To investigate the possibility that bone cell remnants may impose the morphogenetic pattern for bone formation on mesenchymal cells, the following experiments were performed. The demineralized bone matrix was extracted with cold water, or 155 mM sodium chloride for 1–3 days, or Tyrodes solution, or 0.5 M phosphate buffer, or calcium chloride 0.6 M, or urea (0.003–3.0 M), or distilled water at 25°C. These solutions, which rupture cell membranes and extract variable quantities of cytoplasmic substances from small fragments of bone tissue, did not destroy the bone morphogenetic properties of the matrix after implantation in muscle. To investigate the possibility that traces of DNA in remnants of cells in osteocyte lacunae in demineralized cortical bone might impose the

TABLE 2
Yields of New Bone from Matrix Divested of Various Components of Demineralized Matrix

Component	Percent of dry weight of organic matrix	Deactivating, extracting, or digesting agents	Yield of new bone by treated matrix (mg ash/gm)
Cross-linked structure of bone collagen	89.1	High concentrations, H+ Heat > 90 Cryolysis Formalin fixation HNO_2 deamination FDNB, dinitrophenolation IAA, iodoacetamidization ^{60}CO radiation < 2 Mrads UV radiation, 2600–2800 A Acid-alcohol Acid-ether Acid-acetone EDTA at 25°C EGTA at 25°C Pronase Collagenase Lathyrism Penicillamination	0
Uncalcified, cross-linked, osteoid collagen	—	Phosphorus and vitamin D-deficient	0
Noncollagenous acid insoluble and soluble proteins	2.0	$1\ M\ CaCl_2$ $3\ M$ urea Trypsin Elastase	400 ± 50
Glycoproteins	4.6	$1\ M\ CaCl_2$ Hyaluronidase EDTA at 2°C	350 ± 50
Bone sialoprotein	0.4	EDTA at 2°C	300 ± 50
Calcium proteinate	0.001	Citric acid	350 ± 30
Lipid	0.5	Ether Alcohol Acetone Chloroform-ethanol Hexochlorophane Tween 80	400 ± 50
Cell remnants, DNA, RNA, and other cytoplasmic constituents	3.4	$3\ M$ urea DNAase RNAase	300 ± 30

morphogenetic pattern on mesenchymal cells, samples of bone matrix were treated with pancreatic DNase, but the results were positive, as after implantation of bone matrix untreated with DNase (Urist et al., 1968).

To investigate the possibility that RNA might impose the bone morphogenetic pattern on mesenchymal cells, samples were not only incubated in pancreatic RNase, but also irradiated with ultraviolet light, 2600 and 2800 Å, for 24 hours at a distance of approximately 20 cm (420–540 μ W per centimeter of radiation). Enzymatic digestion of RNA did not reduce the quantity of new bone produced by the matrix; UV irradiation of RNA had equivocal effects on yield of new bone (Urist et al., 1968).

Hypothetical Coupling of Collagen Cross-Linkages and the Calcification Initiator to Produce the Bone Morphogenetic Pattern

The experimental observations on deactivated and extracted bone matrix, cited above, associate the bone morphogenetic pattern with the *total* 3-dimensional structure of the organic matrix, free of lipid, and partially depleted of glycoprotein. In an attempt to fit present information into a single hypothesis, Fig. 28 illustrates the relationship between the initial sites of calcification and the fibrous protein cross-linkages in noncalcifying and calcifying tissues. Uncalcified tendon, and osteoid tissue (which does not calcify) lack both calcifiability and the bone morphogenetic property. Lathyritic and penicillaminated matrix (which are calcified but lack the firm cross-links of normal matrix) lack the bone morphogenetic pattern. Only preparations of matrix which were both calcified and normally cross-linked serve as the substratum for bone morphogenesis. The association of the two properties, calcifiability and insolubility in cold dilute HCl, suggest that the nucleation site for calcification and the cross-linked structure of the fibrous proteins form a coupled unit structure. When mesenchymal cells come into close contact with the couple between the apatite nucleation site (elastoid) and protein cross-linked structure, the pattern for bone morphogenesis is imposed on plasma membranes. The quantity of new bone produced by the system is proportional to the mass (Urist et al., 1970) of this coupled structure.

SUMMARY

Demineralized cortical bone matrix is a substratum for bone morphogenesis by postfetal mesenchymal cells. The matrix is demineral-

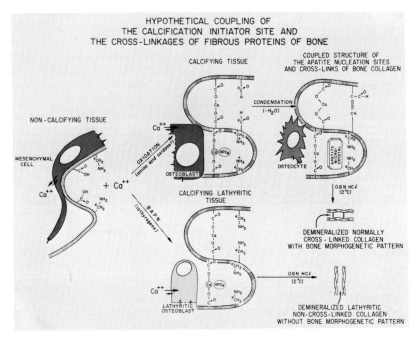

Fig. 28. Diagrammatic representation of the bone morphogenetic pattern in which calcium ions contract the structure of the fibrillar coat-proteins to bring the aldehyde and ϵ-amino groups closer together to interact and produce the cross-linked structure of calcifiable matrix. Tendon collagenous matrix lacks the bone morphogenetic property (left). The bone matrix protein, secreted by osteoblasts (upper right) binds large quantities of calcium ion between adjoining carboxylate groups to produce elastoid and bring ϵ-amino groups into sufficiently close apposition to produce the firm aldehyde cross-link. As phosphate ion association follows at site of uptake of calcium, the nucleation site is formed for crystal formation of apatite. When mesenchymal cell plasmalemma comes into tight *contact* with *calcium-contracted* and firmly cross-linked coupled structure of bone matrix, a chain reaction follows which culminates in bone morphogenesis. The hypothetical calcium cross-link couple is defective in lathyritic matrix (lower right) in which calcium is bound, and the matrix calcifies, but the inhibition of amino acid oxidases by β-amino propionitrile prevents the formation of cross-links even between reactive groups which are in close apposition to one another.

ized in 0.6 N HCl at 2° for 96 hours, and implanted intramuscularly in allogeneic rats. Mesenchymal cells invade old vascular channels of the implant by the 5th day, fuse to form matrix-clasts or differentiate into cartilage by the 10th day, resorb old matrix and differentiate into woven bone by the 15th day, remodel woven bone and differentiate into a central pool of bone marrow by the 20th to 30th days. Correlated observations on the volume of resorbed matrix, differential cell counts, number of mitotic divisions of an average thymidine-^{3}H-labeled cell in the pre- and postosseous phases of bone morphogenesis suggest that some osteoprogenitor cells arise from fission of matrix-clasts. The quantity of woven bone deposited by *progeny* of osteoprogenitor cells is proportional to the *mass* of preimplanted matrix.

As early as the 6th day after implantation, plasma membranes of fixed mesenchymal cells, including filopodial extensions, come into close contact with the substratum of old bone matrix, along surface areas of several square microns, across distances equivalent to the sum of the unit membranes of a desmosome or focal-type tight junction. Coinciding in time with development of substratum-cell contact there is a steep rise in alkaline phosphatase activity which indicates that a bone morphogenetic pattern has been transferred from old matrix to new cell populations.

Experiments on implants of demineralized and undemineralized normal, rachitic, lathyritic, penicillaminated matrix (presented herewith) and observations on bone matrix, extracted chemically, or digested enzymatically, or denatured by irradiation (reported in previous communications) suggest that the total 3-dimensional structure, of the bone matrix, imposes the bone morphogenetic pattern. Preparations of bone matrix, which are lipid-free, low in glycoprotein, divested of all but traces of intracellular material retain morphogenetic properties. Low yield of new bone from implants of rachitic and lathyritic matrix relate the bone morphogenetic pattern to coupling of the properties of calcifiability and insolubility of the fibrous protein. Whether there is similar coupling of these properties in extracellular organic matrix of enamel, dentin, bone, epiphyseal cartilage, aged elastica, placenta and hyaline secretions of urinary bladder epithelium, which also impose a bone morphogenetic pattern on proliferating mesenchymal cells, is presently under investigation.

ACKNOWLEDGMENTS

The writer is indebted to Gilliam L. Molson for the EM studies, and Judith D. Green for the analytical work on rachitic bone matrix, presented in this article.

REFERENCES

BECKER, R. O., and MURRAY, D. G. (1967). A method for producing cellular dedifferentiation by means of very small electrical currents. *Trans. N. Y. Acad. Sci.* **29,** 606–615.

BONUCCI, E. (1967). Fine structure of early cartilage calcification. *J. Ultrastruct. Res.* **20,** 33–50.

BURING, K., and URIST, M. R. (1967a). Transfilter bone induction. *Clin. Orthop.* **54,** 235–242.

BURING, K., and URIST, M. R. (1967b). Effects of ionizing radiation on the bone induction principle in bone implants. *Clin. Orthop.* **55,** 225–245.

DUBUC, F. L., and URIST, M. R. (1967). The accessibility of the bone induction principle in surface decalcified bone implants. *Clin. Orthop.* **55,** 217–223.

ELLISON, M. R., AMBROSE, E. J., and EASTY, G. C. (1969). Chondrogenesis in chick embryo somites *in vitro*. *J. Embryol. Exp. Morphol.* **21,** 331–340.

FELL, H. B. (1969). Role of biological membranes in some skeletal reactions. *Ann. Rheum. Dis.* **28,** 213–227.

FLAXMAN, B. A., REVEL, J. P., and HAY, E. D. (1970). Tight junctions between contact inhibited cells in vitro. *Exp. Cell Res.* **58,** 438–443.

GOODWIN, B. C. (1969). A statistical mechanics of temporal organization in cells. In "Towards A Theoretical Biology" (C. H. Waddington, ed.), p. 140. Aldine Publ., Chicago, Illinois.

GROBSTEIN, C. (1967). Mechanisms of organogenetic tissue interaction. *Nat. Cancer Inst. Monogr.* **2,** 279–299.

HARRIS, H. (1968). "Nucleus and Cytoplasm." Oxford Univ. Press (Clarenden), London and New York.

HOLTFRETER, J. (1968). Mesenchyme and epithelia in inductive and morphogenetic process. In "Epithelial-Mesenchymal Interactions" (R. Fleischmayer and R. E. Billingham, eds.) (*Hahnemann Symp.* 18), pp. 1–29. Williams and Wilkins, Baltimore, Maryland.

HUGGINS, C. B. (1969). Epithelial osteogenesis—a biological chain reaction. *Proc. Amer. Phil. Soc.* **113,** 458–463.

HUGGINS, C. B., and URIST, M. R. (1970). Dentin matrix transformation of alkaline phosphatase and cartilage. *Science* **167,** 896–898.

KONIGSBERG, I. R., and HAUSCHKA, S. D. (1965). Cell and tissue interactions in the reproduction of cell type. In "Reproduction, Molecular, Subcellular, and Cellular" (Michael Locke, ed.), pp. 243–290. Academic Press, New York.

LANE, J. M. (1970). Isolation, characterization and ordering of the CNBr peptides of the α-2 chain of chick bone collagen. *J. Bone Joint Surg.* **52A,** 598.

LASH, J. W. (1968). Somatic mesenchyme and its response to cartilage induction. In "Epithelial-Mesenchymal Interactions" (R. Fleischmayer and R. E. Billingham, eds.) (*Hahnemann Symp.* 18) Williams and Wilkins, Baltimore, Maryland.

MARTIN, G. R., GROSS, J., PIEZ, K. A., and LEWIS, M. S. (1961). On the intramolecular cross-links of collagen in lathyritic rats. *Biochim. Biophys. Acta* **53,** 599.

NANNEY, D. L. (1968). Cortical patterns in cellular morphogenesis. *Science* **160,** 496–502.

NIMNI, M. E. (1968). A defect in intramolecular and intermolecular cross-linking of collagen caused by penicillamine. *J. Biol. Chem.* **242,** 1457–1466.

NOGAMI, H., and URIST, M. R. (1970). A morphogenetic matrix for differentiation of cartilage in vitro. *Proc. Soc. Exptl. Biol.* 530–535.

STRATES, B. S., and URIST, M. R. (1969). The origin of the inductive signal in implants of normal and lathyritic bone matrix. *Clin. Orthop.* **66,** 226–240.

THOMAS, W. C. (1961). Comparative studies on bone matrix and osteoid histochemical techniques. *J. Bone Joint Surg.* **43A,** 419–427.

TRELSTAD, R. L., REVEL, J. P., and HAY, E. D. (1966). Tight junctions between cells in early chick embryo as visualized with the electron microscope. *J. Cell Biol.* **31**, C6–C10.

TRINKAUS, J. P. (1969). "Cells into Organ." Prentiss-Hall, Engelwood Cliffs, New Jersey.

URIST, M. R. (1965). Bone: formation by autoinduction. *Science* **150**, 893–899.

URIST, M. R. (1969). Mesenchymal cell reactions to inductive substrates for bone formation. *In* "Wound Healing" (W. Van Winkle, Jr., ed.). McGraw-Hill, New York.

URIST, M. R. (1970). A morphogenetic matrix for differentiation of bone tissue. *Calc. Tiss. Res.*, Suppl. 4, 98–101.

URIST, M. R., and ADAMS, J. M., JR. (1967). Localization mechanisms of calcification in transplants of aorta. *Ann. Surg.* **166**, 1–18.

URIST, M. R., and CRAVEN, P. L. (1970). Bone cell differentiation in avian species: including comments on multinucleation and morphogenesis. *Fed. Proc., Fed. Amer. Soc. Exp. Biol.* 1680–1693.

URIST, M. R., and DOWELL, T. A. (1968). The inductive substratum for osteogenesis in pellets of particulate bone matrix. *Clin. Orthop.* **61**, 61–78.

URIST, M. R., and NOGAMI, H. (1970). A morphogenetic substratum for differentiation of cartilage in tissue culture. *Nature (London)* **225**, 1051.

URIST, M. R., and STRATES, B. S. (1970). Bone formation in implants of partially and wholly demineralized bone matrix. *Clin. Orthop.* **71**, 271–278.

URIST, M. R., SILVERMAN, B. F., BURING, K., DUBUC, F. L., and ROSENBERG, J. M. (1967). The bone induction principle. *Clin. Orthop.* **53**, 243–283.

URIST, M. R., DOWELL, T. A., HAY, P. H., and STRATES, B. S. (1968). Inductive substrates for bone formation. *Clin. Orthop.* **59**, 59–96.

URIST, M. R., DE LA SIERRA, J., and STRATES, B. S. (1969a). The substratum for new bone formation in tendon. *Clin. Orthop.* **63**, 210–221.

URIST, M. R., HAY, P. H., DUBUC, F. L., and BURING, K. (1969b). Osteogenetic competence. *Clin. Orthop.* **64**, 194–220.

URIST, M. R., JURIST, J. M., JR., DUBUC, F. L., and STRATES, B. S. (1970). Quantitation of new bone formation in intramuscular implants of bone matrix in rabbits. *Clin. Orthop.* **68**, 279–293.

VAN DE PUTTE, K. A., and URIST, M. R. (1966). Osteogenesis in the interior of intramuscular implants of decalcified bone matrix. *Clin. Orthop.* **43**, 257–270.

WEISS, P. (1958). Cell contact. *Int. Rev. Cytol.* **7**, 391–423.

WEISS, P. (1961). Guiding principles in cell locomotion and cell aggregation. *Exp. Cell Res.*, Suppl. 8, 260–281.

YAMADA, T. (1961). A chemical approach to the problem of the organizer. *Advan. Morphol.* **1**, 1–53.

YEOMANS, J. D., and URIST, M. D. (1967). Bone induction by decalcified dentine implanted into oral osseous and muscle tissues. *Arch. Oral Biol.* **12**, 999–1008.

ZWILLING, E., (1968). Morphogenetic phases in development. *Develop. Biol.*, Suppl. 2, 184–207.

Collagen of Embryonic Type in the Vertebrate Eye and its Relation to Carbohydrates and Subunit Structure of Tropocollagen[1]

ZACHARIAS DISCHE

Department of Ophthalmology, College of Physicians & Surgeons, Columbia University, New York, New York 10032

INTRODUCTION TO MORPHOLOGY OF COLLAGEN AND ITS POSSIBLE RELATION TO CARBOHYDRATES

Physical and Chemical Characteristics of Native and Atypical Collagen Fibrils

Collagen is a protein, which for the most part appears in form of nonelastic fibers of varying thickness and considerable tensile strength. In vertebrates this fibrous material fills the intercellular spaces. Its functions are resistance against tensile and shearing forces exerted on various parts of the body and maintenance of the morphological pattern of the organism by providing support to the intercellular matrix in form of tridimensional networks. Because chemical studies on collagen require considerable amounts of material of as high purity as possible they are mainly concerned with highly organized types of collagen, which can be obtained in large quantities from structures such as tendon and skin. In the adult organism collagen appears either in form of fibrils which represent units not further resolvable in the electron microscope without disruptive treatment or bundles of such fibrils of varying thickness. In both cases collagen shows characteristic striation with a period of 600–700 Å in the electron microscope, charactertistic X-ray diffraction patterns, and disappearance of these patterns when collagen contracts at higher temperature.

It is now generally believed that these three physical characteristics of collagen result from specific composition of the peptide chain of collagen in which about one-third of all amino acid residues are represented by glycine and one-fifth by the imino acids, proline and hydroxyproline. Peptides containing these three amino acids form the apolar, crystalline part of the peptide chain. The apolar

---

[1] This work was supported by grants CA-02075 of the National Cancer Institute and EY-00348 of the National Eye Institute.

part alternates with sequences of polar amino acids present to an extent of about 20%. The properties of collagen observed in electron microscopy and X-ray diffraction are found in collagen of most tissues of adult vertebrates. They may however be absent in (a) collagen of some invertebrates like the cuticle collagen of *Lumbricus* and *Ascaris*, (b) collagen of embryos, (c) certain types of collagen in adult vertebrates, e.g., eye tissues, and (d) certain specific types of connective tissue, like basement membranes or laminae basales in adult vertebrates. These atypical forms of collagen can exhibit typical collagen fibrils either during embryonal development or after precipitation from solution after addition of suitable polyanions.

All the above collagens show (with only one known exception) the same characteristic pattern of amino acid composition as typical collagen fibrils in the mature organism. They all contain about 50% of their amino acids in form of glycine and the two imino amino acids and about 20% in form of polar amino acids. The collagen of the tendon and skin of adult vertebrates shares these characteristics with the very different collagen fibrils of the reticulin formation of the kidney and with the collagen of the cuticle of ascaris. The latter is not striated, contains only 2% of the imino acid hydroxyproline (Fitton Jackson, 1960) (Table 1) and shows tropocollagen molecules linked to dimers by S-S bridges (McBride and Harrington, 1967). In biological and particularly developmental studies of collagen, if not in chemical ones, it appears reasonable and useful to use the proportions of glycine, imino acids, and polar amino acids as criterion of molecular identity for collagen.

TABLE 1
Amino Acid Composition of Various Collagens[a]

Amino acids[b]	Human bone collagen	Human tendon extract	Human kidney reticulum	Bovine skin gelatin	Rat skin collagen	*Ascaris* cuticle gelatin
Glycine	319	324	309	334	327	286
Hydroxyproline	100.2	92.1	107.7	92	100	24
Imino acid, total	223.6	218.5	204.9	221	217	304
Acidic amino acid, total	119.2	120.7	129.6	120	121	177
Basic amino acid, total	84.4	84.9	84.4	88.8	84.4	87.4

[a] Compiled from Fitton Jackson (1960).
[b] Residues of amino acids per 1000 residues.

Gross and Kirk (1958) and Fessler (1961a, b) showed that collagen extracted with 0.15 M NaCl at 4°C from skin precipitates at 37°C in the form of native fibrils. Tropocollagen molecules are not stable in pure solution under physiological conditions and aggregate into typical native fibrils. Whenever, therefore, collagen appears in the body of homoiotherms as collagen in solution or as atypical fibrils, an interaction is indicated between the pure collagen protein and some other physiological substances, which either prevent fibrillation or cause appearance of atypical fibrils. Gross et al. (1954) showed that certain polyanions like serum glycoproteins, ATP, acid mucopolysaccharides,[2] and nucleic acids when added to tropocollagen solutions induce the formation of abnormal fibrils (SLS and FLS) with a period of about 2800 A. This fibrillation could be prevented by short-time treatments of tropocollagen solutions with pepsin (Rubin et al., 1963). Kuhn et al. (1961, 1965) showed that this inhibitory effect of pepsin on fibrillation was at least partly due to incomplete removal of pepsin adhering to the collagen molecules. That fibrillation can be affected by substances adhering to collagen molecules under physiological conditions is indicated by alteration of fibrillation of collagen after redissolving SLS fribrils obtained with ATP from tropocollagen solutions (Kuhn et al., 1965). It was suggested (Wood, 1964) that such effects on the fibrillary organization of tropocollagen molecules may be due to complexes of MPS with protein attached to collagen molecules by noncovalent bonds.

Nucleation and Growth of Collagen Fibrils

Physiological polyanions can also regulate the distribution and diameter of collagen fibrils by influencing in a different way the two elementary processes, which take place during fibril formation in solutions of tropocollagen: (1) the nucleation or formation of primary fibrils which form nuclei; (2) the formation of thicker fibrils which arise from the nuclei by a different process of growth by apposition (Wood, 1960a, b; Keech, 1961). These authors showed that chondroitin sulfate A accelerates the nucleation process, but slows down the subsequent growth of the primary fibrils, thus inducing the accumulation of numerous thin fibrils in a tissue. Heparin and DNA, on the other hand, inhibit the process of nucleation.

[2] The following abbreviations are used in this paper: MPS = mucopolysaccharides SO_4GNGl, sulfated sialoglycosaminoglycan; GNCl, guanidine HCl; MEth, mercaptoethanol

K. Meyer suggested (1953) that the acid MPS in the matrix of the connective tissues play a role in the organization of collagen fibrils into units of higher order. Later he related (Meyer 1960) the variations in the composition of the mucopolysaccharide fraction and in the degree of their sulfation in different tissues to differences in the degree of organization of their collagen into fibers of higher order.

The influence of two other types of sugar polymers on the organization of tropocollagen into fibrils and their aggregates of higher order as well as their distribution in tissue matrices was suggested by the author (Dische and Zelmenis, 1955). It has been shown that the thin nonstriated collagen fibrils of the bovine central vitreous are linked to a galactoglucan, which is alkali stable, not associated with hexosamine, and represents about 10% of the protein. The nonorganized collagen protein of the bovine lens capsule had been shown previously to be linked to about 10% of the same kind of heteroglycan (Dische and Borenfreund, 1954). The mature fully organized collagen from various adult tissues had been found to contain no more than 0.5–0.8% of hexose. This hexose constituent in the mature collagen was later identified (Butler and Cunningham, 1966) as a glucogalactoside linked to hydroxylysine. The interfibrillary fluid of the tridimensional fibrillary network of the central bovine vitreous on the other hand was shown to contain, in addition to hyaluronic acid, sialofucohexosaminoglycans bound to protein. One fraction of these hexosaminoglycans, representing about one-third of the total, came out of solution at pH 5 when the hyaluronic acid of the vitreous was broken down by hyaluronidase. The composition of this heteroglycan was different from any so far found in the blood plasma or connective tissue of the ox (Dische et al., 1958). Sialofucoglycoproteins not firmly bound to collagen were also found in bone (Dische et al., 1958) and skin (Boas, 1955). In the reticulin formation of adolescent human kidney, on the other hand, the interfibrillary amorphous matrix was found (Windrum et al., 1955) to contain to an extent of 4–5%, a heteroglycan consisting only of glucose and mannose. No hexose could be found linked to the reticulin fibrils themselves. On the basis of these data, it was suggested (Dische et al., 1958) that the organization of tropocollagen into native fibrils is inhibited by a high content of the hexosamine-free heteroglycan bound to it. The sialofucoglycoproteins in the interfibrillary matrix, as well as the acid MPS would have a role in the spatial distribution of these fibrils and their aggregation into fibril bundles of higher order.

Basis for Higher Level of Organization of Collagen

It is clear that the influence of these different types of carbohydrates and carbohydrate-protein complexes on the morphology of collagen has to be based on covalent or noncovalent bonds between the collagen protein and these carbohydrate compounds. The noncovalent bonds are most likely to be of ionic nature in the case of acid MPS and sulfated glycoproteins. The interaction with sialofucoglycoproteins which are not sulfated can be based on hydrophobic or hydrogen bonds between collagen and constituents of the carbohydrate moiety or the protein moiety of the complex.

But in addition to the interactions between the collagen and carbohydrate components linked to it by covalent bonds or present in the matrix, the structure of the collagen protein itself may influence the organization of collagen on the level of the fibril or fibril aggregates of higher order. There is some evidence that collagen from the same tissue in different species differs in amino acid sequences and the same appears to be the case between collagens from different tissues. Such structural differences may account for variations in the periodicity of native fibrils, the range being between 600 and 700 Å. The hypothesis (Gallop, 1966) that the individual α-chains of tropocollagen are built up of three types of subunits, A, B, and C, appears attractive from this point of view. These subunits are present in varying proportions in various types of collagen and significantly differ in their content in polar amino acids. Collagen disintegrates into these subunits when treated with nucleophilic reagents like hydroxylamine and hydrazine. The molecular weight of subunits analyzed by Gallop was about 17,000, but treatment with nucleophilic reagents can produce subunits of a molecular weight of only about 8000 (Blumenfeld et al., 1965). This degradation of collagen by nucleophilic reagents is apparently due to specific sensitivity of peptide bonds between the α- and β-carboxyls of aspartic acid and the amino group of glycine (Bornstein, 1969). It seems clear that variations in the permutation of such subunits in chains of collagen of different origins can influence the mode of association of tropocollagen molecules in fibrils of different origin.

Differences between Embryonal and Adult Collagen

Electron microscopic studies on connective tissues from developing organisms on fibrogenesis in tissue cultures of chondro- and osteoblasts and on tissue repair processes have revealed three significant characteristics of collagen fibrils in process of development:

1. The diameter of the collagen fibrils which could first be observed during fibrogenesis in cultures of chondro- and osteoblasts was about 80 Å (Fitton Jackson, 1956). It increases during development, but does not exceed about 300 Å during the whole course of embryonal development. Furthermore, the range of the variation of the diameter is much narrower in embryonal than adult tissues (Schwarz, 1960).

2. Embryonal collagen differs in the degree of fibrillary organization. Although in some cases, like that of the chick cornea in the first 5 days of development (Hay and Revel, 1969a), the fibrils with 150 Å diameter show the typical periodicity of about 640 Å, in general one finds in embryonal connective tissues (skin, cartilage) either fibrils with no banding or with a periodicity of only 210 Å (Martin, 1953; Randall, 1954). It is furthermore noticeable that fibrils of different degrees of organization can be present simultaneously at certain stages of the development (Martin, 1953).

3. Embryonal collagen fibrils display a different type of argyrophilia (Schwarz, 1960). The argyrophilic material is found on their peripheral fibers distributed in an irregular pattern. In postembryonal and adult tissues, this material appears concentrated at the site of D bands.

The divergence of embryonal collagen fibrils from the type of organization found in fibrils of the adult organisms could be the result of differences in the primary structure of the embryonal α-chains. Where adult fibrils are formed by apposition of tropocollagen on embryonal fibrils, the latter could still form a core of the adult fibril. This core could have different primary structure, as far as the distribution of polar groups along the peptide chain is concerned. Differences in the distribution of polar groups obviously could lead to changes in the mode of alignment of tropocollagen molecules in fibrils.

Interfibrillar Matrix and Morphogenesis

Components of the interfibrillary matrix could obviously also influence the fibrillary organization by transient or permanent interaction with primordial collagen fibrils. The regulation of the thickness of individual fibrils should mainly depend on this type of interactions. The importance of constituents of the interfibrillary matrix for the morphogenesis of connective tissues manifests itself in differences found in MPS fractions of embryonal pig skin (Meyer *et al.*,

1956a) and rib cartilage (Meyer et al., 1956b; Kaplan and Meyer, 1959). The fact that chondroitin sulfate C appears in conjunction with fine immature collagen fibers, whereas chondroitin sulfate B is mostly associated with fibers in mature tissue points in the same direction. Histochemical differences between embryonal and mature collagen in the localization of the argyrophilic material on the periphery of the fibrils indicate the significance of interactions between collagen and constituents of the matrix. It seems plausible to assume that such interactions are mediated either by electrostatic forces, particularly in the case of acid MPS or by hydrophobic and hydrogen bonds in the case of sialofucoglycoproteins.

Complexes between collagen fibrils and glycosaminoglycans of the interfibrillary matrix, apart from their role in the morphogenesis of the tissue collagen (Wood, 1964), could also be involved in the morphogenetic effects of collagen observed during the development of epithelial organs. Some of the effects, like that on the morphogenesis *in vitro* of embryonal salivary and pancreatic alveoli may be the result of mechanical effects (Grobstein and Cohen, 1965) of collagen fibrils. But the differentiation of skin epidermis of chicken embryos could be achieved in culture only when, between the explanted epithelium and the collagen gel used as support, a basal lamina was formed as a layer of collagen filaments without any periodicity (Dodson, 1967). Some form of nonorganized collagen also appears to play a decisive role in the differentiation process of myoblasts in tissue culture (Koenigsberg and Hauschka, 1966). This suggests the possibility that collagen fibrils of low degree of organization serve as carriers of active molecular groups.

In the case of sulfated acid MPS, variations in number and spatial distribution of SO_4 groups could give rise to di- or tridimensional patterns of ionic forces (Meyer, 1960). In the case of sialofucoglycoproteins, a still greater diversity in the pattern of the spatial distribution of molecular forces could be achieved by variations of the side-chain constituents. One of these constituents, fucose, carries a hydrophobic methyl group. The other, a derivative of neuraminic acid, carries a hydrophilic carbosyl group. The extraordinary variability in the relative number and distribution of fucose and neuraminic acid in relation to species, tissue and functional state of organs (Dische, 1963, 1964) makes them appear particularly suitable as elements of di- or tridimensional patterns of directive forces.

Objectives

To achieve a deeper understanding of the morphogenetic functions of embryonal collagen, it appears necessary to explore the biochemistry of various forms of embryonal collagen and of glycosaminoglycans of the interfibrillary embryonal matrix. Such investigations should not be restricted to the isolation and proper characterization of constituents which are most easily isolated in an adequate state of purity. One should rather attempt to make as complete an inventory as possible, of constituents of the embryonal connective tissue. To obtain adequate amounts of tissue for such investigations, it is necessary in general to use tissue from older embryos. Here the presence of blood vessels, elastic fibers and nerves may complicate the task decisively. This suggests to direct the investigations towards a type of adult tissue, in which the organizational level of collagen and its relation to the ground substance resembles, or is perhaps identical with that of embryonal tissues. The avascular tissues of the eye are of this kind. In the following, therefore, will be reported the results of biochemical investigations concerned with the collagen and the carbohydrates linked to it by covalent or noncovalent bonds in two tissues of the mammalian eye.

NONFIBROUS COLLAGEN OF THE LENS CAPSULE

The lens capsule appears in the electron microscope either as a completely homogeneous structure, as in man, or as consisting of a number of concentric lamellae as in rat or mouse, with no definite appearance of any fibrillary structure. Over 80% of the protein of the capsule which amounts to about 10% of the wet weight consists of a collagen protein. When isolated by gel filtration after pronase digestion of the capsule, this protein has the composition of the typical collagen with 32.5% of the residues represented by glycine and 22.5% by imino amino acids (Kefalides, 1969). The hydroxyproline content in purified collagen was found to be 18% in the rabbit and about 20% in the bovine capsule. Like other collagens, that of the lens capsule is gelatinized by boiling water and 5% trichloroacetic acid at 90°C in 30 minutes.

Saccharides as Solubilizers of Capsular Collagen

The collagen in the lens capsule is apparently kept in solution and protected from fibril formation by the presence of compounds attached to the peptide chain by covalent or noncovalent bonds. Four types of compounds have so far been demonstrated which could be involved in the solubilization of collagen: (1) a disaccharide linked by a glycosidic bond to hydroxylysine residues of the collagen; (2) an acid MPS; (3) a structural complex of a sialoglycosaminoglycan and a protein obtained as a residue insoluble in physiological fluids after the removal of practically all the collagen from the lens capsule by treatment with high purity collagenase; (4) glycoprotein fractions which go into solution with the collagen during the collagenase treatment. The latter have not so far been characterized as molecular units in their original form, but only in form of their degradation products after treatment with pronase (Spiro and Fukushi, 1969).

Disaccharide linked to collagen. The presence of an alkali-resistant sugar polymer consisting of galactose and glucose in a ratio of about 4:3 and free of significant amounts of hexosamine was first demonstrated in the bovine lens capsule (Dische and Borenfreund, 1954) and later in rabbit lens capsules (Dische, 1964). This carbohydrate was later found to be the α-glycosyl β-galactoside of hydroxylysine of the collagen and its amount to represent about 10% of the collagen (Kefalides, 1969; Spiro and Fukushi, 1969). While most of the hexose bound to collagen is represented by this disaccharide, the total amount of galactose found in linkage with collagen exceeds that of glucose.

Acid mucopolysaccharide. On the basis of analytical data obtained in various tissues of animals and man, it is now widely assumed that the acid MPS fraction of the connective tissue of various organisms represents a mixture of different MPS. The proportion of any of these MPS varies to a very considerable degree from one tissue to another. These variations may be related to the organization of the collagen in the intercellular matrix (Meyer, 1960). It appears of particular interest to determine the nature of the hexuronohexosaminoglycans of the lens capsule. A purified preparation of this substance which contained no more than about 10% hexose and protein was obtained by subjecting the lens capsule to a 48-hour digestion at room temperature with a 5% KOH solution in 80% ethanol. This removed all the collagen and left an ethanol-insoluble residue which contained about 60% of the hexuronic acid of the lens capsule (Dische *et al.*, 1967).

The separation of the hexuronoglucosaminoglycan from sialoglycosaminoglycans was then achieved by repeated reprecipitation with 60% acetone at 4°C. Tentative identification of the hexuronide by a combination of two modifications of the carbazole reaction (Dische and Rothschild, 1967) and the orcinol reaction for hexuronic acids (Hoffman et al., 1956; Meyer et al., 1956a) showed that this hexuronide behaves in these reactions like heparitin sulfate obtained from other sources and the material does not contain any significant amounts of hyaluronic acid. Paper chromatography of the hexuronide after hydrolysis according to Dziewiatkowski (1962) in two solvents, confirmed this identification as the R_f of the spots obtained on paper was different from those of heparin and hyaluronic acid and other MPS found in tissues and treated with KOH ethanol like the lens capsule. The optical rotation of this preparation was positive, and almost identical with that reported for heparitin sulfate preparations from other tissues. The preparation contained 1 SO_4 per 5 hexosamines. The presence of additional sulfate groups linked to the amino groups of the glucosamine, however, cannot be excluded, as such sulfate groups would be split off by the treatment with KOH ethanol. The presence of heparitin sulfate in the lens capsule is of particular interest as the lens capsule is produced by epithelial elements and can be regarded as an equivalent of a basal lamina. No hexuronide has been found so far in vascular basement membranes of mesodermal origin.

Sialoglycoproteins. The other type of heteropolysaccharides present in the lens capsule are the sialoglycoproteins or -glycopeptides. Their presence in the lens capsule was first demonstrated in rabbit lens capsules (Dische, 1964) when it was shown that the extraction of the lens capsule with 5% trichloroacetic acid at 90°C for 60 minutes removes all hydroxyproline and about equivalent amounts of galactose and glucose, but only traces of hexosamine and mannose. After further extraction with 10% trichloroacetic acid at 90°C for 6 hours, the amount of mannose in the extract was comparable to that of galactose. The hexosamine which also was present had the elution pattern of glucosamine according to Gardell and R_f of glucosamine in paper chromatography with pyridine-butanol-water as solvent. This fraction also contained significant amounts of fucose. A fraction of very similar composition was later demonstrated in bovine lens capsules by the same procedure and by extraction of the bovine lens capsule with 80% ethanol containing 5% KOH for 48 hours at room temperature and separation of the hexuronide by precipitation with 60% acetone.

a. *Heterogeneity.* Later, more detailed investigations (Kefalides,

1969; Spiro and Fukushi, 1969) led to the preparation of a number of glycopeptides after treatment of bovine and canine lens capsules first with collagenase and then with pronase, gel filtration on Sephadex, and chromatography on Dowex 50. These glycopeptides were probably fragments of larger molecules split by pronase. In these experiments the lens capsules were pretreated by ultrasound to remove all traces of epithelia, and the possibility does not seem to have been excluded that some of the glycoproteins may have gone into solution during the ultrasound treatment. The heterogeneity of the sialoglycoprotein component of the lens capsule was demonstrated by preparing in our laboratory without the use of any other enzyme than high purity collagenase, a complex of a sialofucoglucosaminoglycan with a protein. This fraction was practically free of collagen and contained only a fraction of the total hexosamine of the sialoglycoprotein of the capsule (Rothschild and Dische, 1970).

b. *Preparation and purification of a structural glycoprotein fraction.* The complex was obtained by incubating bovine lens capsules in four times their volume of $\frac{1}{50}$ M Tris buffer pH 7.4 containing 0.005 M $CaCl_2$ and 3.4 mg per 10 ml of high purity collagenase (Worthington) and four antibiotics, which by microbiological tests were shown to suppress bacterial growth in the material. The incubation was carried out for 16 hours at room temperature under shaking. The digest was centrifuged and the residue was washed, then dried with acetone; the same extraction procedure was repeated four times until no significant amounts of hydroxyproline were extracted. A small residue remained which was insoluble in water, saline, 8 M urea, 5% trichloroacetic acid, and 1% sodium dodecyl sulfate. It consisted of about 10% carbohydrate and 90% protein. The carbohydrate contained about equal amounts of galactose, mannose and glucose, and glucosamine. The ratio of hexosamine to hexose was 0.4. It was characterized by a very high content of fucose. The molar ratio fucose:hexosamine was 0.35. After 2-hour treatment at 37°C in 0.01 M HCl containing per milliliter 2 mg of crystalline pepsin, the material went into solution. When incubated with 0.01 M β-mercaptoethanol (MEth) for 16 hours at room temperature, most of the complex went into solution, leaving only a small residue. After dialysis of the supernatant the composition of its carbohydrate was the same as in the original material and the content in protein was somewhat smaller. The fractionation with acetone of this solution yielded one fraction which precipitated between 40 and 50% acetone and a second fraction precipitating at 80% acetone. These two fractions accounted for all the carbohydrate and for 90% of the protein. A carbohydrate-free protein residue corresponding to 10% remained in the supernatant acetone. The 50% acetone fraction showed a higher content in protein and a great part of it became insoluble after precipitation but could be resolubilized by reincubation with 0.01 M MEth. This indicates that reducing groups in this fraction underwent reoxidation under the influence of air and possibly some catalytic agent in the fraction. The 80% acetone fraction remained completely soluble and could be reprecipitated without a change in its composition by 80% ethanol. All these data indicate that the insoluble residue of the collagenase treatment of the lens capsule represents a complex of a protein with the characteristic sialofucoglucosaminoglycan. This complex, however, represents only a fraction of the total sialofucoglucosaminoglycans present in the lens capsule which, so far, could not be prepared in purified form but were only obtained as glycopeptides after fragmentation by pronase (Kefalides, 1969; Spiro and Fukushi, 1969).

Role of S-S linkages in the stabilization of the complexes of capsular collagen and glycoproteins. Experimental data indicate that the stability of the lens capsule, as well as of the glomerular basement membranes depends upon the presence of disulfide linkages (Spiro, 1969). When these are reduced in presence of 8 M urea, one part of the collagen and noncollagen proteins of the lens capsule and glomerular basement membrane goes into solution (Kefalides and Winzler, 1966; Kefalides, 1970; Spiro, 1969, 1970). The fact that the structural complex of sialoglycosaminoglycan, (GNGl) and protein which remains as an insoluble residue after exhaustive collagenase treatment of the lens capsule can be solubilized by incubation with β-MEth suggested to test the possibility of removing this structural complex from the lens capsule by incubation of the latter with β-MEth. Incubation with 0.1 M β-MEth will not bring any hexosamine-containing material into solution. However, when the lens capsule was incubated with β-MEth and 5 M GNCl, practically all of the structural complex went into solution.

To test what effect the removal of this structural complex would have on the collagen of the lens capsule, the following experiment was carried out: Four hundred bovine lens capsules were divided into three equal parts. One part served as control. One part (I) was incubated in 5 M GNCl solution which had a pH 4. The second (II) in 5 M GNCl solution adjusted to pH 7. The extraction period was 16 hours at room temperature with shaking. The extraction mixtures were centrifuged, and the supernatants were removed and replaced by the same volume of the respective GNCl solutions. This procedure was repeated three times, and it was shown that there was no significant extraction of either carbohydrate or collagen during the fourth extraction period. The extraction residue of sample I was then divided into two equal parts. From one half, guanidine was removed by exhaustive dialysis of the residue. This half of sample I served for the determination of the effect of GNCl alone on sialoglycoproteins and collagen. The other half of sample I and the extraction residue of sample II were suspended in 5 M GNCl solution of pH 4 and pH 7, respectively, and β-MEth was added up to a concentration of 0.1 M to both samples. They were then incubated for three successive periods of 16 hours at room temperature, the solution of β-MEth in 5 M GNCl being renewed after every extraction period of 16 hours. The supernatants of the β-MEth extractions and the extraction residues of both halves of sample I and of II were then exhaustively dialyzed. The dialyzed residues from the two halves of sample I, sample II and of the control sample were subjected to an exhaustive digestion with collagenase in $M/50$ Tris buffer and 0.005 M $CaCl_2$ containing 3.4 mg of collagenase in 10 ml of Tris buffer. The residues of all four samples were analyzed. These analyses showed that in both samples I and II treated with MEth in 5 M GNCl, more than 90% of the structural carbohydrate protein complex went into solution. In the one half of sample I which was treated only with GNCl pH 4, but not β-MEth, only 30% of the structural complex went into solution. The effect of the extraction with GNCl alone at pH 4 and of GNCl plus β-MEth at pH 4 and 7 on the collagen of the lens capsule was then investigated. To this end, hydroxyproline was determined in all dialyzed extracts of both halves of sample I and of II and in all collagenase digests of residues of these three samples and of the control sample. This determination showed that in the sample first extracted with GNCl pH4 and then extracted with guanidine and MEth pH4, in two experiments, 19 and 40%, respectively, of the collagen was apparently lost during dialysis. In this extraction 78% of the total collagen of the lens capsule was solubilized, whereas the extraction with guanidine at pH 7 and MEth solubilized only 27% of collagen and the extraction with GNCl at pH 4 alone solubilized no more than 12 to 13% of collagen. From these experiments, two conclusions can be drawn:

1. The structural glycosaminoglycan–protein complex is bound by noncovalent links to the collagen and cannot form part of one single protein chain together with collagen peptides.

2. Up to 40% of the collagen of the lens capsule must be present in the form of polypeptide chains with a molecular weight not larger than 10,000 linked either to each other, or to lenticular glycoproteins but not by peptide bonds usually found in proteins. These bonds must rather be either noncovalent or S-S linkages or peptide bonds of some unknown kind, so labile that they are disrupted by 5 M GNCl at pH 4.

Control experiments in our laboratory in which acid-soluble collagen from calf skin was treated with 5 M GNCl pH 4 and by the same guanidine solution in combination with 0.1 M β-MEth under identical conditions as in the experiments with lens capsule showed no loss of collagen during the exhaustive dialysis of the extracts. Therefore, the presence of such short chains of collagen protein with a carbohydrate content of about 10%, or of some specific, unusually labile peptide bonds of unknown nature between short chain, carbohydrate rich, collagen polypeptides appears to be characteristic for the lens capsule. It is furthermore clear that S-S linkages play [as was claimed by Spiro (1968) for glomerular basement membranes] a significant role in the stabilization of the capsular structure. They link collagen peptides either to each other or to glycoproteins, or as constituents of the protein moiety of glycoproteins, they create conditions necessary for noncovalent bonding between collagen units and glycoproteins.

Developmental Age of Lens Capsule in Relation to Collagen and Glycosaminoglycans

In the growing animal the levels of MPS and sialoglycoproteins in the lens capsule reach their maxima at a much earlier stage than does collagen. This was first demonstrated on rabbit capsules (Dische and Zelmenis, 1965). Hydroxyproline, nitrogen, hexosamine, and hexuronic acids were determined in lens capsules of rabbits between the ages of 2 weeks and one year. The capsules were successively extracted 30 and 60 minutes with 5% trichloroacetic acid, and for additional 17 hours with 10% trichloroacetic acid at 90°C. These determinations showed that whereas hexosamine and hexuronic acids were at a maximum at an age of 4 weeks, hydroxyproline was at that time at a level only half as high as in 3-month to 1-year-old animals. These data suggest that heparitin sulfate and at least one part of sialoglycosaminoglycans of the lens capsule have an influence either on the synthesis of the capsular collagen, or on its organization, or both.

CORNEAL STROMA: CONNECTIVE TISSUE ARRESTED AT AN EMBRYONAL STAGE OF DEVELOPMENT

The bovine corneal stroma appears in the electron microscope to consist of layers of collagen fibrils with a typical periodicity of about 640 Å in orthogonal arrangement. The diameter of the fibrils varies in a narrow range between 230 Å and 300 Å. It was reported (Schwarz, 1953) that the type of argyrophilia of the corneal fibrils of the adult animal is the same as that found in embryonal collagen fibrils. Furthermore, the diameter of the fibrils and their distribution in the individual layers of the stroma, reach their final stage during the embryonal development of the animal and remain unchanged during the adult life. This seems to be an essential factor in assuring the transparency of the cornea. The collagen fibrils of the corneal stroma appear, therefore, on the basis of the three criteria for the structural characteristics of embryonal collagen fibrils, to be of embryonal type.

During the embryonal development of the chick, collagen fibrils with the typical periodicity of about 640 Å are formed in the cornea during the first 5 days of embryonal life. They appear in orthogonal arrangement at a time when mesenchymal cells do not yet penetrate the space between the corneal epithelium and the lens (Hay and Revel, 1969b). They appear, therefore, to be of epithelial origin and are replaced after the fifth day of the development by fibrils formed by mesenchymal cells. As the collagen fibrils grow in diameter by apposition of new tropocollagen, the possibility cannot be precluded that the epithelial fibrils remain at least in some of the stromal fibrils as an inner core. Whether the development of the stroma in mammals follows the same course as that in chick has not yet been established by adequate electron microscopic studies. The cornea, therefore, offers an opportunity to study the role of the stromal carbohydrates in the maintenance of the embryonal character of the collagen fibrils and of their characteristic orthogonal arrangement which appears also in other forms of embryonal connective tissue of ectodermal origin.

As in the case of the lens capsule and other forms of connective tissue, several types of carbohydrate are found in the corneal stroma.

One is a glucogalactan attached to collagen fibrils themselves, the others are polysaccharides present in the interfibrillary matrix of the stroma.

Carbohydrate-Rich Extractable Collagen of the Stroma

It had been reported (Woodin, 1952) that when corneal stroma is exhaustively extracted with 1 M NaCl, the extraction of corneal MPS does not proceed as far as is the case when 1 M $CaCl_2$ at pH 8 is used for the extraction. This suggested that the stromal MPS are partly present in the interfibrillary matrix in free form and partly linked to each other or to other constituents of the matrix or are distributed among different morphological elements of the matrix. In our preliminary experiments, the extraction of MPS by 1 M $CaCl_2$ at pH 8 was accompanied by an extraction of a collagen protein which was insoluble in distilled water (Dische and Robert, 1963). This suggested to investigate in which way the nature of the salt solution used for MPS extraction influenced the extraction of this soluble collagen protein.

Extraction. Two sets of experiments were carried out in which the extraction of MPS and collagen, and its course in 0.16 M NaCl at pH about 6 was compared with that by 1 M $CaCl_2$ at pH 8. The corneal stroma was left to swell overnight in saline at 4°C and then sliced with a hand slicer into about an equal anterior, medium, and posterior part. The posterior part was not investigated because of possible complications from the presence of Descemet's membrane. The stroma was homogenized in a VirTis homogenizer with ice cooling for three 1-minute intervals separated from each other by a 1-minute cooling off period. In one set of experiments (A) the stroma was then extracted five times at 4°C with five times its volume of 0.16 NaCl for 16 hours. After each 16-hour period, a supernatant was obtained by centrifugation, the residue was resuspended in the same volume of saline, and the extraction was repeated. The fifth extraction with saline was followed by five or six extractions with 1 M $CaCl_2$ pH 8 carried out under identical conditions. In the second set of experiments (B) the procedure was identical except that each individual extraction was followed by a homogenization for 3 minutes carried out the same way as the initial homogenization at the beginning of the experimental set A. The supernatant from each extraction was exhaustively dialyzed, first against running, then against distilled, water. The precipitates which appeared in the retentate were collected and the collagen in them solubilized either by 5% trichloroacetic acid at 90°C in 60 minutes or by exhaustive degradation with high purity collagenase (Worthington). The dialyzed supernatants were concentrated to a suitable volume by lyophilization. Hydroxyproline, hexosamine, and hexuronic acid were determined in each sample. Hexoses were determined by a procedure described elsewhere (Dische, 1970) using two modifications of the cysteine H_2SO_4 reactions of hexoses (Dische and Danilchenko, 1967) under the assumption that the hexose linked to hexosamine is present in form of keratosulfate. The subtraction of this hexose from the total hexose yielded the disaccharide linked to collagen. In two experiments of type A the extraction of MPS and of collagen decreased sharply with each following extraction of 0.16 M NaCl and became barely significant in the fifth extraction. When the $CaCl_2$ extraction followed the extraction with saline, the concentration of collagen in the first $CaCl_2$ extraction was about ten times as high as in the last NaCl extraction and dropped in the following extractions almost to zero. The ratio of the total collagen extracted by $CaCl_2$ to that extracted by NaCl was about the same as the ratio of hexuronic acid extracted by the two salts (Table 2). In the B series experiments, however, in which the stroma was homogenized between each extraction, the course of extraction of collagen was different. Here significant amounts of collagen went into solution not only in the first $CaCl_2$, but also in the following ones, although there was a continuous drop in the amount of collagen extracted with each subsequent extract to a constant very low level. The ratio of the total amount of collagen extracted by $CaCl_2$ to that extracted by NaCl was in the B series significantly higher than the ratio of the hexuronic acid extracted by the two salt solutions. (Table 2) The parallelism between the amounts of the extractable collagen and MPS obtained with the two different salt solutions in series A experiments and the increase in this collagen observed after repeated homogenization of the tissue (series B), indicates that the extractable collagen is present in at least two different structures. This conclusion appears supported by results of the following third series of extraction experiments. The stroma residue from the series A extraction was further extracted for three consecutive 48-hour periods at 4°C with shaking with a 5 M GNCl solution of pH 4. An amount of collagen several times as large as that present in the last $CaCl_2$ extract went into solution during the first extraction

TABLE 2

Comparison of the Amounts of Hexuronic Acid and Hydroxyproline Extracted by 0.16 NaCl and 1 M CaCl$_2$ pH 8, Respectively, from the Corneal Stroma Separated from the Descemet Membrane

	A[a, c]				B[b, c]		
Exp.	Extraction	Hexuronic acid	Hydroxy-proline	Exp.	Extraction	Hexuronic acid	Hydroxy-proline
I.	a. NaCl	7.85	1.62	III.	a. NaCl	12.3	3.85
	b. CaCl$_2$	3.48	1.10		b. CaCl$_2$	3.2	3.56
II.	a. NaCl	2.76	1.03	IV.	a. NaCl	9.8	10.5
	b. CaCl$_2$	7.10	2.55		b. CaCl$_2$	4.3	16.0

[a] Stroma homogenized only before the first NaCl extraction.
[b] Homogenized before each single extraction.
[c] Values are expressed as milligrams per 50 corneas.

period. Much less was found in the second guanidine extract and almost none in the third extract. This indicates a limited pool of the collagen material which remains in the stroma after extraction in series A. The ratio of hexose:collagen in these fractions was very similar to that found in the collagen extracted by CaCl$_2$ in B series (Dische, 1970). The collagen extracted in series A seems to be present in the interfibrillary matrix. Such a localization is suggested first by the parallelism between its extraction and that of MPS by NaCl and CaCl$_2$ solutions. Furthermore, it is suggested by the fact that the proportion of extractable collagen of the total collagen is several times as high in calf corneas as in adult ones. It appears improbable that calf fibrils should contain a much larger quantity of extractable collagen than fibrils of older animals as the structural conditions of the fibrils is an important factor in the transparency of the cornea. The collagen which goes into solution only under the mechanical stress of repeated homogenization or treatment with GNCl could be derived from the material of different electron density distributed irregularly along the fibril and removable by ultrasonication (Schwarz, 1953).

Content and composition of the carbohydrate. The determination of the carbohydrate, free of hexosamine and linked to the extractable collagen fraction, shows values between 2% and 3% (Dische, 1970). This value is about three times as high as that found for collagen-linked hexose in the main bulk of the stromal collagen and other collagens. It was shown by paper chromatography and extinction coefficients in the two forms of the cysteine H_2SO_4 reaction of hexoses, that this carbohydrate consists of about equal amounts of galactose and glucose (Table 3).

The total amount of the extractable collagen is about 4.5% of the total collagen of the cornea. This was calculated on the basis of its hydroxyproline content by multiplying this value by 7.46.

Hypothesis about the structural significance of the extractable collagen. At least one part of the carbohydrate-rich extractable stromal collagen seems to accompany the MPS constituents of the matrix and to be present in the interfibrillary matrix in an amorphous form. The collagen may form together with the MPS a gel filling the interfibrillary space. Collagen gels have been shown to induce the differentiation of the epithelial and embryonic muscle cells in tissue culture (Dodson, 1967; Koenigsberg and Hauschka, 1966). During the development of the corneal stroma these collagen gels in the matrix could either influence the differentiation of keractocytes or directly affect the orthogonal arrangement of the collagen fibers of the stroma. Such an inductive function could be visualized, for instance, in the following way: The gel could consist of a tridimensional network of tropocollagen molecules in a hexagonal arrangement. The interaction between the tropocollagen molecules of the network and the collagen of the fibrils could be different at the crossing points of the tropocollagen fibrils forming the hexagonal network

TABLE 3

RATIOS OF HEXOSE TO COLLAGEN IN DIALYZATES FROM COLLAGENASE DIGESTS OF THE THREE COLLAGEN FRACTIONS SOLUBILIZED BY 0.15 NaCl, 1 M CaCl$_2$, AND 5 M GUANIDINE HCl AND OF THE INSOLUBLE COLLAGEN AFTER CaCl$_2$ EXTRACTION

Collagen	Amount of collagen (mg)	Hexose[a] / Collagen
1. NaCl extracted	10.8	2.0
2. CaCl$_2$ extracted	7.0	2.1
3. Insoluble residue after CaCl$_2$ extraction	135.0	1.1
4. Guanidine extracted	10.3	3.6

[a] Hexose calculated on the basis of galactose:glucose = 1 except for guanidine extract of 1:2. Collagen = hydroxyproline × 7.46.

and in the center of each hexagonal space. The collagen fibrils have a diameter of about 340 Å in the hydrated state and are separated from each other by the same space interval. One-half milligram of collagen per cubic centimeter of the stroma would be sufficient to form a hexagonal network in which each side of each cube would be about 1400 Å long. Such an interaction of the tropococollagen of the gel with collagen fibrils could be mediated by polar groups of MPS molecules aligned along individual tropocollagen molecules, thus forming a tridimensional field of ionic forces which could also contribute to the orthogonal arrangement of the corneal fibrils.[3]

The high content in hexose of the extractable collagen could serve either to prevent fibril formation by this tropocollagen and thus facilitate the formation of a tridimensional network on a molecular level, or it could facilitate the formation of complexes between collagen and hexosaminoglycans. There is no evidence so far of such complexes between collagen and MPS molecules. On the other hand, it had been shown that in addition to MPS, the corneal stroma contains a sulfated sialoglycosaminoglycan (SO_4GNGl) (Robert and Dische, 1963) which is not extracted by those salt solutions which almost quantitatively extract the MPS fractions. This suggested that SO_4GNGl may be fixed to the collagen of the corneal stroma and, thus play the role of a structural constituent of the stroma. This problem has been studied in our laboratory, as well as by Robert and his co-workers (1963, 1965) for the last few years.

Complex between a Sulfated Sialoglycosaminoglycan and a Carbohydrate-Rich Collagenpolypeptide

Preparation and Analysis. Our basic experiments were carried out in such a way that corneal stroma from which MPS were removed by salt extraction were exhaustively treated for 16 hours at room temperature with high purity collagenase (Worthington) in $\frac{1}{50}$ M Tris buffer containing 0.005 M $CaCl_2$. The treatment with collagenase at room temperature was repeated until no significant amounts of hydroxyproline could be found in the collagenase containing extraction fluid (Table 4). A combination of four antibiotics or $CHCl_3$ was

---

[3] Such a tridimensional network would be analogous on a molecular scale with the tridimensional, approximately hexagonal network of thin nonstriated collagen fibrils of the central part of the vitreous body. Here the collagen fibrils have a diameter of 100–250 Å and the interfibrillary space is filled with a gel of hyaluronic acid and sialoglycoproteins.

TABLE 4
Composition of Fractions of the NaCl and CaCl₂ Extractable Collagen (A) and the Nonextractable Stroma (B) Obtained by Successive Incubation with Collagenase for 16 Hours at Room Temperature

Extraction No.	Hydroxyproline[a]		Hexose as galactose[a]		Hexosamine[a]	
	A	B	A	B	A	B
1	770	11,100	390	1740	48.5	14.5
2	12	1900	135	750	32	50
3	2	1740	69	240	14.5	40.5
4	1.5	22	45	132	8.2	36
5	0	5	45	112	8.2	36
6	0	0	45	97	8.2	35
Residue	10.7	90	80	270	16.7	60

[a] Values are expressed as micrograms in 10 ml of extract.

used to prevent bacterial growth. The antibiotics used were shown by microbiological tests to prevent bacterial growth in the cornea. After such exhaustive collagenase treatment, a small residue remained which consisted of a protein containing about 6% hydroxyproline and a sialofucoglycosaminoglycan (Table 4). The composition of the latter was determined by digesting this residue with 80% ethanol containing 5% KOH at room temperature for 48 hours and dissolving the residue of this digest, which was completely free of hydroxyproline, in water. The analysis of the carbohydrate and paper chromatography with pyridine-butanol-water as solvent, showed that it consisted of galactose and mannose in about equivalent amounts, glucosamine, and a small amount of fucose and neuraminic acid. In addition, it contained 1 molecule of SO_4 for each hexosamine. This carbohydrate, therefore, was obviously identical with the SO_4 GNGl present in the residue from the extraction by 5% trichloroacetic acid at 90°C of the bovine corneal stroma (Robert and Dische, 1963). The determination of the composition of the purified SO_4GNGl permitted to calculate the amount of hexose constituents bound to hexosamine in the total residue from the collagenase treatment of the stroma. By subtracting this hexose from the total hexose in the residue, the amount of the hexose linked to collagen was obtained. The amount of collagen protein in the residue was found by determining hydroxyproline and N_2 in two subsequent extracts of the residue each obtained by heating at 90°C for 30 minutes in 5% trichloroacetic acid. Assuming that the hydroxyproline:collagen ratio and the small amounts of noncollagen protein were identical in both fractional extracts, the amount of collagen in the residue could be calculated. As this procedure yielded the amount of noncollagen protein extracted in 60 minutes by 5% trichloroacetic acid at 90°C, it made it also possible to determine the amino acid composition of the collagen residue. This was done by subtracting from the amount of each amino acid present in the extract the amount derived from the noncollagen protein. The latter was calculated from the amino acid composition of the residue of the two 30-minutes trichloroacetic acid extractions, which did not contain any collagen any more. This determination showed that the collagen of the residue contained 325 residues of glycine, 36 residues of hydroxylysine, and 205 imino acid residues of which 41 were hydroxyproline. It satisfies, therefore, the compositional criterion for collagen protein used in the present discussion. By subtracting the hexose of the SO_4GNGl bound to hexosamine from the hexose of the total residue, it was shown that the collagen in the residue contained 10% hexose. It consisted of glucose and galactose. It was obvious, therefore, that $SO_4 \cdot$ GNGl has a specific affinity to a collagen or a fragment of a collagen chain particularly rich in carbohydrate. That SO_4GNGl is bound by noncovalent bonds to the carbohydrate-rich collagen chain could be shown in the following way.

The corneal stroma freed of MPS by NaCl and $CaCl_2$ extraction was exhaustively extracted by repeated treatment with 5 M GNCl of pH 4 at 4°C with shaking. Three different extraction periods of 48 hours each were used. When the corneal stroma after extraction with GNCl was freed of guanidine by dialysis and subjected to digestion with collagenase simultaneously with a control sample of the stroma which was freed of MPS by NaCl and $CaCl_2$ extraction but was not treated with GNCl, the ratio of hexose to hydroxyproline was significantly higher in the collagenase digest of the guanidine-treated stroma than in the control stroma. On the other hand, the residue remaining after exhaustive collagenase extraction of the GNCl-treated sample contained only a small fraction of the SO_4-GNGl and the hexose-rich collagen linked to it found in the control stroma not treated with GNCl (Table 5). These results show clearly that the treatment with GNCl of the MPS free corneal stroma removes the SO_4GNGl which after exhaustive collagenase treatment of the corneal stroma not treated with GNCl remains combined with

TABLE 5
EFFECT OF THE EXTRACTION OF THE TOTAL INSOLUBLE STROMA
COLLAGEN BY 5 M GUANIDINE HCL ON THE COLLAGENASE-
RESISTANT COLLAGEN–GLYCOPROTEIN COMPLEX

Sample	Hydroxy-proline (μg/cornea)	Hexosa-mine (μg/cornea)	Collagen hexose (μg/cornea)	Collagen hexose as % of collagen	
				A[a]	B[b]
I. Stroma collagen before further treatment	141	52	188	8.4	1.07
II. Stroma collagen extracted with 5 M guanidine HCl	7	12	23.9	21.4	1.30

[a] In collagenase-resistant residue.
[b] In first collagenase digest of the stroma.

the carbohydrate rich collagen as a collagenase-resistant residue. This resistance to collagenase is eliminated when SO_4GNGl is removed from the residue by the treatment with 5 M GNCl. The digestion with collagenase of the guanidine-treated stroma brings, therefore, into solution the carbohydrate-rich collagen residue. The latter is not digested by collagenase in the control stroma because of its linkage to the structural SO_4GNGl by noncovalent bonds.

The carbohydrate-rich collagen linked to SO_4GNGl could represent a separate molecular unit present in small amounts in the stroma or it could be a part of a chain of a stromal collagen which is particularly rich in carbohydrate and is split off during the collagenase digestion from the α-chain. It would not go into solution because of its link with the structural SO_4GNGl. It has recently been suggested (Gallop, 1966) that the α-chain of collagen consists of three types of subunits with a MW of 17,000 and which differ in their amino acid composition. The main bulk of all the hydroxylysine appears in only one of these subunits. It seems conceivable that the hydroxylysine in glycosidic linkage with the hexose is concentrated even in a smaller part of the chain of collagen. If the mature collagen contains six hexose residues and these residues are all concentrated in a part of the peptide chain which represents one-tenth of the total α-chain, then the hexose content of this subunit of a molecular weight of about 10,000 would be about 10%. This is an agreement with the content in hexose found in the collagenase resistant collagen residue linked to the structural SO_4GNGl. It should be noted that it was possible by nucleophilic agents to obtain a considerable portion of the α-chain in form of subunits of MW only of 8000 (Blumenfeld et al., 1965).

Presence of short-chain collagenpolypeptides. The presence of short chains in the carbohydrate-rich collagen linked to structural SO_4GNGl could be demonstrated by incubating the complex of the collagen with the structural SO_4GNGl for 16 hours at room temperature in 0.01 M β-MEth. When the supernatant from this treatment was dialyzed, the retinant contained most of the hexosaminoglycan but only traces of hydroxyproline. The residue of the MEth extraction, however, contained only a fraction of the original amount of hydroxyproline present in the complex before treatment with MEth (Table 6). It is clear, therefore, that some hydroxyproline-containing peptide chains were removed by MEth together with the structural SO_4GNGl in a dialyzable form. They are, therefore, present already in the collagenase-resistant collagen residue as molecules with a M.W. of about 10,000 or less. That the collagenase-resistant collagen residue is split off from the stromal collagen by collagenase is also indicated by the fact that collagen extracted from the stroma together with MPS by 0.16 M NaCl and $CaCl_2$ also leaves, after exhaustive collagenase treatment, a carbohydrate-rich collagen residue linked to a sialoglycosaminoglycan. This finding can be barely reconciled with the assumption that the carbohydrate-rich, collagenase-resistant collagen represents a separate molecular unit comparable in size with the α-chain.

Hypothesis about the significance of the complex for the organiza-

TABLE 6
Composition of the Collagenase-Resistant Residue of the Stroma before and after Treatment with Mercaptoethanol (MEth)

Sample	Hydroxy-proline (μg per cornea)	Hexo-samine (μg per cornea)	Collagen hexose[a]
I. Residue before MEth treatment	141	52	188 (8.4%)
II. Residue I extracted with MEth			
a. Fraction dissolved in MEth	10	31	26 (16.1%)
b. Fraction not dissolved in MEth	63	12.4	98 (9.8%)

[a] Values in parentheses represent the content of collagen in hexose in each fraction.

tion of stromal collagen fibrils. All the above-mentioned data suggest that the SO_4GNGl of the corneal stroma has a specific affinity to one part of the α-chain of the stromal collagen which is characterized by a very high content in hydroxylysine and the glycosylgalactan linked to it. If one considers the quantitative relation between the amounts of SO_4GNGl and the collagen of the stroma, it seems possible to visualize a function of SO_4GNGl in the orthogonal arrangement of the collagen fibrils of the stroma. The total amount of hexosamine linked to hexose in the corneal stroma from which the MPS has been removed shows that there is about 2.7 mg of hexosamine for 1 gm of collagen. SO_4GNGl contains five molecules of hexose for each three molecules of hexosamine, the ratio of fucose and of neuraminic acid to hexosamine is about 0.05, and there is one SO_4 for one hexosamine. There would be, therefore, 8.3 mg of hexosamine containing carbohydrate per gram of collagen. If we assume that the corneal collagen fibrils contain tropocollagen molecules in a hexagonal arrangement, then the surface layer of the fibril should contain about one-third of the total collagen present in the fibrils. If we further assume that each tropocollagen molecule on the fibril surface is linked to one molecule of a glycoprotein consisting of SO_4GNGl and a protein moiety, then we have to assume a M.W. of about 7500 for the sulfated glycan itself. If one assumes that SO_4GNGl links always one tropocollagen molecule on the surface of a fibril of one lamella with another tropocollagen molecule on the surface of a fibril in the neighboring lamella, then the length of the molecule of the connecting glycoprotein should correspond to 340 Å, the distance between collagen fibrils in hydrated form. If we further assume that the glycoprotein connects also two points on two fibrils in adjacent lamellae, both situated laterally with respect to the main axis of the fibrils the M.W. for the total glycoprotein complex has to be between 25,000 and 50,000. The ratio of carbohydrate to protein in the glycoprotein would then be between 3, 5, and 7. All these values are of the order of magnitude usually found in glycoproteins. The quantitative relation, therefore, between SO_4GNGl of the corneal stroma and the collagen is in agreement with the concept that the sulfated stromal glycoprotein forms a tridimensional network which keeps the collagen fibrils of the stroma in their orthogonal arrangement. It should be noted that such a network would not preclude the ability of the stroma to swell as the two linkages of the glycoprotein with two tropocollagen molecules situated on neighboring fibrils must not be identical and one of them could be very labile and be disrupted by the hydration pressure of the corneal stroma. The rigidity of the interfibrillary gel and the negative charges could keep the sulfated glycoprotein chain in extended position so that subsequent dehydration of the stroma could lead to a restitution of the original structural conditions. If this concept of the role of the structural SO_4GNGl in the orthogonal organization of the corneal stroma has merit, then the enzymatic degradation of the structural glycoprotein, for instance, by trypsin should lead to a collapse of the orthogonal arrangement under some mechanical stress or chemical treatment. Trypsinization experiments on the primary corneal stroma of the chick embryo (Hay and Revel, 1969b) seem to confirm this assumption. Electron microscopy of the primary stroma treated with trypsin shows clearly the dissolution of the orthogonal arrangement which is, on the other hand, maintained after treatment with hyaluronidase (Hay, personal communication).

BIOLOGICAL IMPLICATIONS

The presence in the corneal stroma of a carbohydrate-rich collagen-polypeptide, with particular affinity to the structural sulfated sialoglycoprotein of the cornea which apparently represents a segment of chain split off by collagenase, suggests that such short-chain constituents of the α-chain may play a major role in the inductive interactions between collagen and epithelial formations during the embryonal development. In this role it would function as a carrier of active molecules which would bring the latter in contact with various elements of the developing organism. For such a function, a certain mobility of the carrier would be of importance. The subunit concept of the α-chain as it emerges from recent investigations seems to provide a basis for such a mobility.

Although there is at present no conclusive evidence that the subunits of α-chains are linked to each other by other than peptide bonds, some of the linkages must be of particular lability as they are not only split by nucleophilic reagents, but apparently also by mechanical stress in presence of guanidine. Thus, Miller *et al.* (1967) found that extraction of soluble collagen from the bone of young chicks yields some molecules with a M.W. of only 50,000. Recent observations in our laboratory (unpublished) show that soluble collagen can be broken down by mechanical stress in concentrated salt solutions into still smaller subunits. Tropocollagen subunits synthetized in embryonic osteo- and chondroblasts could be obtained as separated subfractions by gradient centrifugation of homogenates of such cells (Fitton Jackson, 1965). The same author also found that the heaviest of these fractions assembled seemingly spontaneously under the influence of centrifugal forces to larger units which could form typical collagen fibrils. This was related to their nature as subunits of the α-chain.

We report in this paper that incubation of bovine lens capsule in 5 M GNCl at pH 4 in 0.01 M MEth solubilizes completely the structural complex of a sialoglycosaminoglycan and a protein and simultaneously brings into solution up to 40% of the capsular collagen as carbohydrate-rich short collagen chains which are dialyzable.

Spiro (1970) has shown that large molecular unit obtained from bovine glomerular basement membranes by pronase treatment and gel filtration disaggregate in β-MEth and 8 M urea into very short chains of collagen protein.

These data strongly suggest that short-chain collagen polypeptides rich in disaccharide and linked by noncovalent bonds to active glycoprotein are building blocks of laminae basales. These complexes could be involved as carriers of active groups in the inductive influences of laminae basales on the differentiation of various epithelia and myoblasts in tissue cultures.

SUMMARY

Differences in the morphology on the electron microscopic level between collagen fibrils of adult and embryonal connective tissue are discussed. A role of saccharide polymers attached to fibrils or present in the interfibrillary matrix as causative factors in these differences is suggested. The similarity between the morphology of the embryonal collagen and that of certain connective tissues of the vertebrate eye is pointed out. The presence of three types of saccharide polymers in the lens capsule, considered as an example of a morphological structure containing only unorganized collagen, is reported and their composition is described. The preparation and analysis of a structural complex between a sialoglycosaminoglycan particularly rich in fucose, and a protein is described. This insoluble complex is solubilized by incubation in 0.01 M β-mercaptoethanol. The incubation of the capsule in 5 M guanidine HCl and 0.1 M β-mercaptoethanol at pH 4 at room temperature brings into solution with the structural complex most of the collagen, up to half of the latter in form of short dialyzable chains. This suggests noncovalent or S-S bonds between the structural glycoprotein and short collagen chains, which stabilize the latter within the capsule.

The extraction and preparation of a carbohydrate-rich collagen from the corneal stroma is reported. Its localization in the interfibrillary matrix is suggested on the basis of the parallelism in the extraction conditions between mucopolysaccharides and this collagen and its increased concentration in corneas of young animals. It is suggested that it forms part of an interfibrillary gel and a factor in the developmental organization of the corneal stroma.

The preparation of a collagenase-resistant complex between a sulfated sialoglycoprotein and a collagen polypeptide containing about 10% carbohydrate is reported. The glycoprotein can be detached by 5 M guanidine at pH 4 from the collagen protein. Incubation with 0.1 M β-mercaptoethanol brings into solution the glycoprotein and one part of the collagen moiety of the complex. The latter is detached in the form of short dialyzable chains. Some evidence indicates that the latter are carbohydrate-rich subunits split off from the stromal collagen with high affinity toward the sulfated structural glycoprotein. On the basis of this affinity and quantitative relations between the sulfated glycoprotein and total stromal collagen, the hypothesis is set forth that the sulfated glycoprotein plays a role as structural factor in orthogonal arrangement of the corneal fibrils. The carbohydrate rich subunits of the collagen may play a role as carriers of active proteins in inductive developmental processes.

The dynamic influence of collagen in developmental processes is, according to this hypothesis, due to the presence of specific subunits in the α-chains. These subunits are linked to each other or other amino acids either by peptide bonds of particular lability or noncovalent bonds or S-S linkages. They are, furthermore, characterized by a high content of a disaccharide linked glycosidically to hydroxylysine and by high affinity to active glycoproteins. They could also be present in the tissue as individual short-chain units or easily detached from larger units.

REFERENCES

BLUMENFELD, O. O., ROJKIND, M., and GALLOP, P. M. (1965). Subunits of hydroxylamine-treated tropocollagen. *Biochemistry* **4**, 1780.

BOAS, N. F. (1955). Distribution of hexosamine in electrophoretically separated extracts of rat connective tissue. *Arch. Biochem. Biophys.* **57**, 367.

BORNSTEIN, P. (1969). The nature of hydroxylamine sensitive bond in collagen. *Biochem. Biophys. Res. Commun.* **36**, 957.

BUTLER, W. T., and CUNNINGHAM, L. W. (1966). Evidence for the linkage of a disaccharide to hydroxylysine in tropocollagen. *J. Biol. Chem.* **241**, 3882.

DISCHE, Z. (1963). Chemical specificity of glycoproteins of animal tissues. *Exposes Ann. Chem. Med.* p. 49.

DISCHE, Z. (1964). The glycans of the mammalian lens capsule—a model of basement membranes in small vessel involvement in diabetes mellitus (M. D. Siperstein et al., eds.) *Amer. Inst. Biol. Sci. Publ.* p. 201.

DISCHE, Z. (1967). The informational potentials of conjugated proteins. *Protides Biol. Fluids. Proc. Collog.* **14**, 5.

DISCHE, Z. (1970). The carbohydrate rich collagen of the corneal stroma. "Chemistry and Molecular Biology of the Intercellular Matrix"(E. A. Balasz, ed.), Vol. 1 p. 229. Academic Press, New York.

DISCHE, Z., and BORENFREUND, E. (1954). The polysaccharide of the lens capsule and its topical distribution. *Amer. J. Ophthalmol.* **38**, 165.

DISCHE, Z., and DANILCHENKO, A. (1967). Modifications of two color reactions of hexoses with cysteine and sulfuric acid. *Anal Biochem.* **21**, 119.

DISCHE, Z., and ROBERT, L. (1963). The carbohydrates of the corneal stroma. *Fed. Proc., Fed. Amer. Soc. Exp. Biol.* **21**, 172.

DISCHE, Z., and ROTHSCHILD, C. (1967). Two modifications of the carbazole reaction of hexuronic acids for the differentiation of polyuronides. *Anal. Biochem.* **21**, 125.

DISCHE, Z., and ZELMENIS, G. (1955). Polysaccharides of the vitreous body. *Arch. Ophthalmol.* **54**, 528.

DISCHE, Z., and ZELMENIS, G. (1965). The content and structural characteristics of the collagenous protein of rabbit lens capsules at different ages. *Invest. Ophthalmol.* **4**, 174.

DISCHE, Z., DANILCHENKO, A., and ZELMENIS, G.(1958). The neutral heteropolysaccharides in connective tissue. *Chem. Biol. Mucopolysaccharides Ciba Found. Symp. 1957* p. 116.

DISCHE, Z., ZELMANIS, G., and ROTHSCHILD, C. (1967). The hexosaminohexuronide of the bovine lens capsule. *Arch. Biochem. Biophys.* **121**, 685.

DODSON, J. W. (1967). The differentiation of epidermis. The interrelationship and dermis in embryonic chicken skin. *J. Embryol. Exp. Morphol.* **17**, 83.

DZIEWIATKOWSKI, D. D. (1962). Separation of hexuronic acids by ion exchange process. *Biochim. Biophys. Acta.* **56**, 167.

FESSLER, J. H. (1961a). Some properties of neutral salt soluble collagen. I. *Biochem. J.* **76**, 452.

FESSLER, J. H. (1961b). Properties of neutral salt soluble collagen. II. *Biochem. J.* **76**, 463.

FITTON Jackson, S. (1956). The morphogenesis of avian tendon. *Proc. Roy. Soc. Ser. B* **144**, 556.

FITTON JACKSON, S. (1960). Connective tissue cells. *In* "The Cell" (J. Brachet and A. E. Mirsky, eds.), Vol. 6, p. 426.

FITTON JACKSON, S. (1965). Antecedent phases in matrix formation. *In* "Structure and Function of Connective and Skeletal Tissue" (S. Fitton Jackson et al., eds.), p. 277. Butterworth, London.

GALLOP, P. M. (1966). 3, 2, 1: A B C subunit hypothesis for the chains of tropocollagen. *Nature (London)* **209**, 73–74.

GROBSTEIN, C., and COHEN, J. (1965). Collagenase: Effect on the morphogenesis of embryonic salivary epithelium in vitro. *Science* **150**, 628.

GROSS, J., and KIRK, K. D. (1958). The heat precipitation of collagen from neutral salt solutions. Some rate regulating factors. *J. Biol. Chem.* **233**, 355.

GROSS, J., HIGHBERGER, J. H., and SCHMITT, F. O. (1954). Collagen structures considered as states of aggregation of a kinetic unit. The tropocollagen particle. *Proc. Nat. Acad. Sci. U.S.* **40**, 679–692.

HAY, E. D., and REVEL, J. P. (1969a). In "Fine Structure of the Developing Avian Cornea," p. 32. Karger, Basel.

HAY, E. D., and REVEL, J. P. (1969b). In "Fine Structure of the Developing Avian Cornea," p. 39. Karger, Basel.

HODGE, A. J., HIGHBERGER, J. H., DEFFNER, G. G., and SCHMITT, F. O. (1960). The effects of proteases on the tropocollagen macromolecule and on its aggregation properties. *Proc. Nat. Acad. Sci. U.S.* **46**, 197–206.

KAPLAN, D., and MEYER, K. (1959). Ageing of human cartilage. *Nature (London)* **183**, 1267.

KEECH, M. K. (1961). The formation of fibrils from collagen solutions IV. Effect of mucopolysaccharides and nucleic acids. An electron microscope study. *J. Biochem. Biophys. Cytol.* **9**, 193.

KEFALIDES, N. A. (1969). Characterization of the collagen from lens capsule and glomerular membranes. *Proc. Congr. Int. Diabetes Fed. 6th, Stockholm, 1967,* p. 307. Excerpta Med. Found., Amsterdam, 1969.

KEFALIDES, N. A. (1970). In "Biochemistry of Basement Membranes" (A. E. Balasz, ed.), Vol. 1, p. 535. Academic Press, New York.

KEFALIDES, N. C., and WINZLER, R. J. (1966). The chemistry of glomerular basement membrane and its relation to collagen. *Biochemistry* **5**, 702.

KOENIGSBERG, I. R., and HAUSCHKA, S. D. (1966). The influence of collagen on the development of muscle clones. *Proc. Nat. Acad. Sci. U.S.* **55**, 119–126.

KUHN, K., KUHN, J., and HANNIG, K. (1961). Einwirkung von Trypsin auf gelostes. Kollagen. *Hoppe Seylers Z. Physiol. Chem.* **326**, 50.

KUHN, K., SCHUPPLER, G., and KUHN, J. (1965). Alteration in the behaviour of collagen with progressive purification and after treatment with proteolytic enzymes. In "Structure and Function of Connective and Skeletal Tissue" (S. Fitton Jackson, et al. eds.), p. 64. Butterworth, London.

LOEWI, G., and MEYER, K. (1958). The acid mucopolysaccharides of embryonic skin. *Biochim. Biophys. Acta* **27**, 453.

MCBRIDE, O. W., and HARRINGTON, W. F. (1967). Ascaris cuticle collagen; on the disulfide cross linkages and the molecular properties of the subunits. *Biochemistry* **5**, 1484.

MARTIN, A. V. W. (1953). Fine structure of cartilage matrix. *Nature Struct. Collagen, Papers Disc.* p. 129.

MEYER, K. (1953). Hyaluronic acid, chondroitin sulfates and their protein complexes. *Disc. Faraday Soc.* **13**, 271.

MEYER, K. (1960). Nature and function of mucopolysaccharides of connective tissue. In "Molecular Biology" (D. Nachmansohn, ed.), p. 69. Academic Press, New York.

MEYER, K., DAVIDSON, E., LINKER, A., and HOFFMAN, P. (1956a). The acid mucopolysaccharides of connective tissue. *Biochim. Biophys. Acta* **21**, 506.

MEYER, K., HOFFMAN, P., und LINKER, A. (1956b). Mucopolysaccharides of costal cartilage. *Science* **128**, 896.

MILLER, F. J., MARTIN, G. R., PIER, K. A., and POWERS, M. (1967). Characterization of chick bone collagen and compositional changes associated with age. *J. Biol. Chem.* **242,** 5481.

RANDALL, J. T. (1954). Observations on the collagen system. *Nature (London)* **174,** 853.

ROBERT, B., PARLEBAS, J., and ROBERT, L. (1963). Etude immunochimique d'une glycoprotéine de la cornée. *C. R. Acad. Sci.* **256,** 323.

ROBERT, L., and DISCHE, Z. (1963). Analysis of a sulfated sialofucoglucoasminogalactomannosidoglycan from corneal stroma. *Biochem. Biophys. Res. Commun.* **10,** 209.

ROBERT, L., and PARLEBAS, J. (1965). Biosynthèse in vitro des glycoprotéines de la cornée. *Bull. Soc. Chim. Biol.* **47,** 1853.

ROTHSCHILD, C., and DISCHE, Z. (1970). Insoluble complex of non collagen protein and sialoglycosaminoglycan from mammalian lens capsule. *Fed. Proc., Fed. Amer. Soc. Ex. Biol.* p. 922.

RUBIN, A. L., PFAHL, D., SPEAKMAN, P. T., DAVISON, P. E., and SCHMITT, F. O. (1963). Tropocollagen. Significance of protease-induced alterations. *Science* **139,** 37.

RUBIN, A. L., DRAKE, M. P., DAVISON, P. F., PFAHL, D., SPEAKMAN, P. T., and SCHMITT, F. O. (1965). Effects of pepsin treatment on the interaction properties of tropocollagen macromolecules. *Biochemistry* **4,** 181.

SCHWARZ, W. (1953). Untersuchungen über die Differenzierung der Cornea und Sklera Fibrillen des Menschen. *Z. Zellforsch. Mikrosk. Anat.* **38,** 78.

SCHWARZ, W. (1960). Heutige Vorstellungen über die Ultramikroskopische Struktur des Bindegewebes in Struktur und Stoffwechsel des Bindegewebes. W. H. G. Hauss, ed.), p. 108 Thieme, Stuttgart.

SPIRO, R. G. (1969). Chemistry of the renal glomerular basement membrane. *Proc. Congr. Int. Diabetes Fed. 6th Stockholm, 1967,* p. 586. Excerpta Med. Found., Amsterdam.

SPIRO, R. G. (1970). Biochemistry of basement membranes. *In* "Chemistry and Molecular Biology of the Intercellular Matrix (A. E. Balasz, ed.), Vol. 1, p. 516. Academic Press, New York.

SPIRO, R. G., and FUKUSHI, S. (1969). The lens capsule. Studies on the carbohydrate units. *J. Biol. Chem.* **244,** 2049.

WINDRUM, G. M., KENT, P. W., and FASTOE, J. E. (1955). The constitution of human renal reticulin. *Brit. J. Exp. Pathol.* **36,** 49.

WOOD, G. C. (1960a). The formation of fibrils from collagen solution. I. The effect of experimental conditions. Kinetic and electron microscope studies. *Biochem. J.* **75,** 588.

WOOD, G. C. (1960b). The formation of fibrils from collagen solutions. II. A mechanism of collagen fibril formation. *Biochem. J.* **75,** 598.

WOOD, G. C. (1964). The precipitation of collagen fibers from solution. *Int. Rev. Connect. Tissue Res.* **2,** 1–31.

WOOD, G. C., and KEECH, M. G. (1960). The formation of fibrils from collagen solutions, Effect of chondroitin sulphate and some other naturally occurring polyanions on the rate of formation. *Biochem. J.* **75,** 588–597.

WOODLIN, A. M. (1952). The corneal mucopolysaccharide. *Biochem. J.* **51,** 319–330.

Intra- and Extracellular Control of Epithelial Morphogenesis

MERTON R. BERNFIELD AND NORMAN K. WESSELLS

Lt. Joseph P. Kennedy, Jr. Laboratories for Molecular Medicine, Department of Pediatrics, and Department of Biological Sciences, Stanford University, Stanford, California 94305

INTRODUCTION

The major developmental processes taking place during embryonic organ formation are cytodifferentiation and morphogenesis. Cytodifferentiation involves cellular specialization and occurs by the selective loss and acquisition of specific functions. Morphogenesis involves the arrangements of cell populations in a precise and organized fashion. These processes occur in an ordered, sequential manner and ultimately result in an organ morphology essential to the function of the differentiated cells. The concept that the process of cellular specialization produces unique cell types is familiar, but less emphasis has been given to the fact that the morphogenesis of each organ is as unique as is its pattern of cytodifferentiation. Since there is a mutual dependence between the formation of unique structures and the acquisition of specific functions, mechanistic distinctions between the two processes are imprecise. A more meaningful distinction may be to view the developmental history of each organ as including a "determination event," and a subsequent period of organ-specific gene action. The early, morphogenetic events are dictated by certain genes, and other genes are activated later to direct the ontogeny of the final cellular phenotype. Each of these processes—determination, morphogenesis and cytodifferentiation—has been attributed to the class of interactions between dissimilar cell populations known as "embryonic induction." Analysis of the intra- and extracellular bases of morphogenesis has permitted a more incisive understanding of such inductions. The purpose of this paper is to discuss some recent findings in our laboratories which provide insights into these bases of morphogenesis.

Organ morphogenesis includes four types of cellular behavior: movement of individual cells (as in spinal ganglion formation from neural crest cells; Weston, 1970); changes in shape of cell groups (as in neural tube formation from medullary plate; Baker and Schroeder, 1967); differential mitosis (as in condensation of early limb mesenchyme; Janners and Searls, 1968); and differential cell death (as in sculpturing of digits; Saunders, 1966). In any single organ, for example the salivary gland of a rodent embryo, only a few of these properties may be involved. Since organogenesis encompasses both the actions of the component cells and of the cells as a group, we will present studies relating to individual cell participation in morphogenesis, to the way in which these actions are integrated and to the means by which the resultant changes in organ form are controlled.

EXTRACELLULAR CONTROL OF MORPHOGENESIS

Morphogenesis of a variety of organs involves interactions between epithelial primordia and their investing mesenchyme. In these epitheliomesenchymal interactions, the development of characteristic epithelial morphology may require the presence of mesenchyme at specific times. Some epithelia require homologous mesenchyme (e.g., salivary and bronchial epithelium) for characteristic development; others will develop normally under the influence of mesenchyme from several sources (e.g., pancreatic or skin epithelium) (Golosow and Grobstein, 1962; Dodson, 1967) or even in the absence of mesenchyme if culture media are supplemented with embryo extracts (Rutter et al, 1964; Ronzio and Rutter, 1969). Morphogenesis may involve progressive branching (as in the cases of salivary, lung, or ureteric bud epithelia), or the cells may assume histotypic patterns (e.g., acinar arrangements in the pancreas, basal cell orientation and proliferation in the skin). In several instances, the characteristic morphogenesis of an epithelium is modified by the mesenchyme with which it is grown, producing an epithelium which reflects the origin of the mesenchyme (Auerbach, 1960; Alescio and Cassini, 1962; Taderera, 1967; Dameron, 1968; Kratochwil, 1969; Spooner and Wessells, 1970b). This effect of mesenchyme in controlling morphogenesis is poorly understood, although numerous explanations have been advanced (Grobstein, 1967; Holtfreter, 1968; Wolff, 1968). Epithelial morphogenesis can occur under conditions where the epithelium and mesenchyme are separated by a porous filter (Grobstein, 1967; for review). These studies have unequivocally demonstrated that direct mesenchymal contact is not required for the interaction and have focused attention on the role of extracellular materials in morphogenesis.

Collagen is the most abundant extracellular protein of vertebrates. A variety of observations suggest that collagen plays an important role in morphogenesis (Fitton Jackson, 1968; for review). Ultrastructural studies of the epitheliomesenchymal interface frequently reveal collagen-like fibers (Edds and Sweeney, 1961; Hay and Revel, 1963; Kallman and Grobstein, 1964; and others), and this collagen is thought to be related to the development of characteristic morphology (Stuart and Moscona, 1967; Wessells and Evans, 1968b). Abundant collagen fibrils in a highly ordered array are present in the basement lamella underlying the basement membrane of the epidermis of aquatic vertebrates. The precise organization of this collagen has been extensively studied, and recent work by Berliner (1969) and Nadol et al. (1969) has provided evidence that the epidermis is the source of this characteristic orientation. The ordering of the collagen fibrils depends upon the continued presence of the epidermis and may involve the secretion of glucosamine-labeled material by the epidermal cells (Berliner, 1969). Nadol and co-workers (1969) implicated the epidermal basement membrane in the initiation of collagen fibril assembly and the establishment of fibrillar organization.

Collagen associated with epithelia is thought to arise from mesenchyme (Kallman and Grobstein, 1965; Bernfield, 1970), although in certain cases, the epithelial component may contribute collagenous proteins (Hay and Revel, 1969; Goldberg and Green, 1968). Epithelia cultured transfilter from mesenchyme possess collagen upon their surfaces that is apparently derived from the transport of collagenous proteins across the filter. This transfilter passage of collagen from mesenchyme is not dependent upon epithelial morphogenesis (Bernfield, 1970). Indeed, the epithelium probably regulates the amount of collagen synthesized by mesenchyme (Berliner, 1969; Bernfield, 1970).

Studies utilizing collagenase on transfilter cultures suggest that collagen may play a role in epithelial morphogenesis. Collagenase treatment of salivary, lung, or ureteric bud epithelia grown transfilter from mesenchyme causes a loss of the normal branched epithelial morphology and a temporary cessation of epithelial morphogenesis (Grobstein and Cohen, 1965; Wessells and Cohen, 1968). Pancreatic epithelia are unaffected by similar enzyme treatment. The probable site of action of collagenase is upon the basement membrane and its associated collagen. Electron microscopic studies

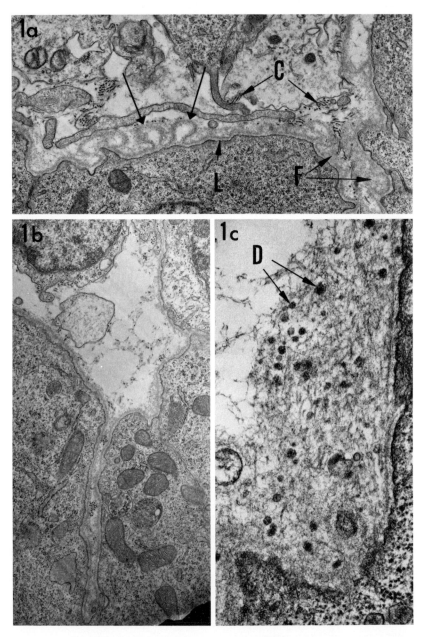

FIG. 1a. The epitheliomesenchymal junction of a normal 13-day embryonic salivary gland showing typical extracellular materials. Collagen (*C*), basal lamina (*L*), and fine filamentous materials (*F*) are seen. This view is of cells near the base of a large cleft in the surface of the epithelium, and shows the unusual convolutions of basal lamina (arrows) found only in such regions. × 12,500.

of collagenase-treated rudiments show disruption of the basement membrane and adjacent extracellular matrix.

Although collagen associated with epithelial surfaces may be important for stabilizing the shape of developing epithelia, other extracellular substances may be equally important. The epitheliomesenchymal junction includes the epithelial basement membrane and its associated extracellular materials (Fig. 1). These interfacial components are closely adherent to the surface of the epithelium and ultrastructurally consist of a pleomorphic three-layered structure (Kallman, et al., 1967; Wessells, 1968; Rifkind, et al., 1969). The layer immediately adjacent to the epithelial cell plasma membrane is an electron lucent zone of uniform thickness which may represent the glycocalyx or cell surface coat (Ito, 1969). The middle layer, the basal lamina, is electron opaque and less uniform. External to this dense band is material of varying thickness and structure. Fibers with a periodicity characteristic of collagen fibers traverse this superficial layer. Investigations by Rambourg and Leblond (Rambourg et al., 1966; Rambourg and Leblond, 1967) suggest that all three layers may represent the periodic acid–Schiff (PAS) positive material seen histochemically at the epitheliomesenchymal interface of many embryonic glands (Slavkin and Bavetta, 1968; Grobstein, 1967). Virtually all this material is removed by treatment with crude trypsin-pancreatin during the process of separating intact epithelium from mesenchyme (Grobstein, 1967; Kallman et al., 1967, Kallman and Grobstein, 1966). PAS-positive material (Dodson, 1967) and the basal lamina of the basement membrane (Kallman and Grobstein, 1966) are replaced even in the absence of mesenchyme. These observations support the idea that the epithelial basement membrane is predominantly of epithelial cell origin (Pierce, 1966).

The chemical nature of the basement membrane has been carefully studied in adult tissues (renal glomeruli, anterior lens capsule) and is thought to consist of three major components, a large molecular weight glycoprotein with high carbohydrate content, a smaller

---

FIG. 1b. A narrow cleft in the surface of a cultured salivary epithelium. Every cleft observed contains extracellular materials similar to those seen here. × 12,500.

FIG. 1c. A high power view showing various salivary gland extracellular materials. The peculiar dense bodies (D) have been observed in developing salivary gland, lung, pancreas, skin, feathers, and hairs, at stages in which epitheliomesenchymal interactions are thought to be occurring. The composition or origin of the dense bodies is unknown. Note that those bodies lack membranes and, therefore, are not extensions of cell surfaces (as in Slavkin et al., 1969b). × 38,000.

glycoprotein with lower carbohydrate content, and a very specialized form of collagen with a high content of largely glycosylated hydroxylysine residues (Kefalides and Winzler, 1966; Spiro, 1967; Kefalides, 1968; Spiro and Fukushi, 1969). Although very few chemical studies of embryonic basement membrane have been performed, the epitheliomesenchymal interface of embryonic tissues characteristically demonstrates PAS-staining material and an electron dense basal lamina, both characteristic of adult epithelial basement membranes.

Ontogenetic evidence suggests a role for the basement membrane in morphogenesis. Chick epidermis stripped from dermis with EDTA maintains its basal cell orientation when cultured in contact with a collagen gel or dead dermis and produces a band of PAS-staining material at the base of the epithelial surface (Dodson, 1967). However, trypsin-separated epidermis in the presence of high concentrations of embryo extract will also exhibit normal basal cell behavior apparently in the absence of a basement membrane (Kallman et al., 1967). The collagenase treatment of various epithelial rudiments (see above) removes the basement membrane as determined ultrastructurally, but reappearance of a dense basal lamina is associated with resumption of normal morphogenesis (Wessells and Cohen, 1968). The PAS-positive material interposed between the epithelial and mesenchymal components of the developing tooth provides a suitable substrate upon which both epithelial and mesenchymal cells align, and form a columnar cell layer (Slavkin, et al., 1969b).

These data are consistent with the hypothesis that interfacial materials are supportive of morphogenesis. The nature of the essential material in the extracellular matrix is unknown. Although collagen has been frequently implicated, the evidence is indirect and circumstantial. For example, pancreatic epithelia do not undergo branching morphogenesis when grown transfilter from mesenchyme and do not undergo a change in shape when treated with collagenase. Ultrastructural studies demonstrate that although normal pancreatic epithelia have collagen at their surfaces, normal acinar differentiation can occur under experimental conditions (Kallman and Grobstein, 1964) in which *no* collagen accumulates near the epithelium. In the absence of embryo extracts, collagen in the form of a heat or ammonia-polymerized gel is not sufficient in itself to support pancreatic development (Wessells and Cohen, 1966), or to support epidermal basal cell orientation and mitosis (Wessells, 1964). Collagen may be essential only in cases where epithelial branching is the predominate morphogenetic pattern. Wessells and Cohen (1968) noted a correlation between collagenase sensitivity and specific mesenchymal requirement, and suggested that organ-specific morphogenesis may be dependent upon the distribution of collagen at unique and specific epithelial sites. Collagen is distributed on lung epithelium in greatest amounts at morphogenetically quiescent areas (trachea, interbronchial regions) and in least amount at the morphogenetically most active sites [bronchial tips (Wessells, 1970)]. Similar distributions characterize salivary stalk and interlobular clefts *versus* the surfaces of growing and branching lobules (Kallman, unpublished; Grobstein and Cohen, 1965). Salivary morphogenesis does not occur in response to nonsalivary mesenchyme which deposits collagen on the epithelium (Bernfield, 1970). Although the site of collagen distribution on the epithelium may be critical, collagen per se does not appear to be a sufficient or active cause for the development of certain characteristic epithelial structures.

Three major pieces of evidence led us to investigations of the role of surface-associated mucopolysaccharides in morphogenesis. Kallman and Grobstein (1966) showed that the epithelium synthesizes hyaluronidase-susceptible surface-associated material that labels with glucosamine-^{3}H. In addition, crude trypsin-pancreatin treatment removes the epithelial basement membrane and, in the studies described below, causes a loss of characteristic epithelial contour. Moreover, studies of purified collagenase (see below) revealed that commercial preparations of clostridial collagenase, as well as of a highly purified collagenase preparation, contain appreciable mucopolysaccharidase activity. Consequently, the enzymes used in the studies implicating the importance of collagenase-susceptible materials in morphogenesis undoubtedly contained mucopolysaccharidase activity, thus casting doubt on the interpretation of these data.

The submandibular salivary gland of the mouse embryo was used to search for morphogenetically active surface-associated materials since much information on this gland is available (Borghese, 1950; Grobstein, 1953), and because of the following advantages: (1) The submandibular glands of 13¼ day mouse embryos are multilobed, yet have not developed to the point where complete removal of mesenchyme is diffidult. (2) Studies of acid mucopolysaccharides are feasible, since these glands are not sufficiently differentiated to produce the mucins and sulfated glycoproteins characteristic of the mature gland. (3) The intact embryonic glands undergo character-

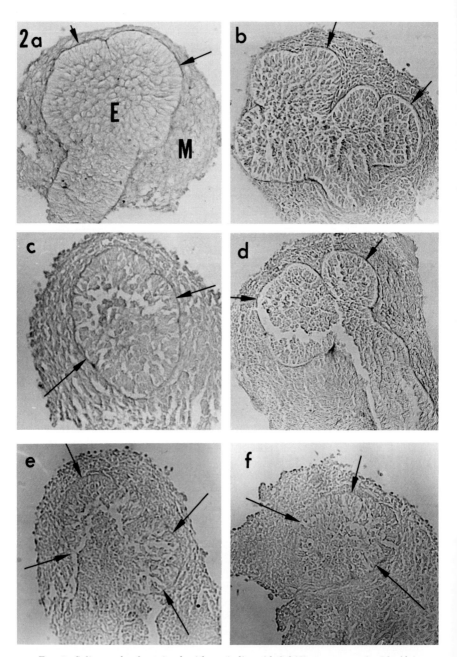

FIG. 2. Salivary glands stained with periodic acid–Schiff reagents and with Alcian Blue at various magnesium concentrations. × 175. (a) The periodic-acid Schiff reagents stain the junction (arrows) between the epithelium (E) and the mesenchyme (M). (b–f) Alcian Blue 8 GX stains the interface (arrows) in the presence of 0.2 M (b),

istic morphogenesis *in vitro*, so that the initial 3-6 lobes branch progressively and form multiple lobules with characteristic acini. At 13¼ days, the epithelium may be separated intact from the mesenchyme by various enzymatic treatments in conjunction with mechanical agitation. (4) When these isolated epithelia are recombined with fresh mesenchyme and placed in organ culture, the mesenchyme rapidly surrounds the epithelium, reconstituting the original relationship between epithelium and mesenchyme.

ACID MUCOPOLYSACCHARIDE AT THE EPITHELIOMESENCHYMAL JUNCTION

Acid mucopolysaccharides are distinguished from glycoproteins and neutral polysaccharides by (1) a high proportion of hexosamines, (2) polyanionic properties due to sulfate ester groups and/or carboxyl groups of uronic acids, (3) the unique presence of uronic acid residues, and (4) susceptibility to digestion by β-1 → 4-hexosaminidases. Because of difficulties in unequivocally demonstrating acid mucopolysaccharides histochemically (Quintarelli, 1968), all these features were expoited in an attempt to show that acid mucopolysaccharide exists at the epitheliomesenchymal junction of developing salivary glands (Bernfield and Banerjee, 1971a).

Staining of the Epitheliomesenchymal Junction

Periodic acid-Schiff (PAS). Neutral polysaccharides and glycoproteins are strongly reactive with PAS reagents, because the basis of such staining involves oxidation of vicinal hydroxyl groups in sugars to aldehydes, which then react with Schiff reagents to form a colored complex. The staining reaction has little specificity, but has been used to indicate the presence of epithelial basement membranes. Acid mucopolysaccharides are not PAS-reactive when tissues are stained in the usual manner (Lillie, 1965). Whole salivary rudiments stained with PAS demonstrate appreciable staining material at the epitheliomesenchymal junction (Fig. 2a). The stain is uniformly distributed around the surface of the epithelium.

Alcian Blue. Alcian Blue is a polyvalent cationic dye that binds with polyanions by predominantly electrostatic linkages (Scott et al., 1964). Numerous workers have shown that complexation and precipitation of polyanions by organic cations is reversed in the

---

0.3 M (c) and 0.4 M (d) MgCl$_2$. Staining of the interface (arrows) disappears almost completely at 0.6 M (e) and 0.7 M (f) MgCl$_2$, indicating that the Mg^{2+} concentration at which the dye is displaced from the interface is between 0.4 and 0.6 M.

presence of sufficient concentrations of inorganic electrolyte. Scott and co-workers (Quintarelli et al., 1964; Scott and Dorling, 1965) have carefully investigated the polyanion–Alcian Blue complexation and have demonstrated that magnesium ion will displace the dye from acid mucopolysaccharides in solution and in tissue sections. The concentration of Mg^{2+} required to displace the dye ("critical concentration") depends upon the type of anionic group, the frequency of charged groups, and the molecular weight of the polysaccharide. At pH 5.8, complications arising from masking of polysaccharide by protein are minimized (Scott and Dorling, 1965). In the absence of electrolyte, staining is nonspecific, but the "critical electrolyte concentration" can be found by determining the concentration of Mg^{2+} at which staining of a specific polyanion is reversed.

Thus, using Mg^{2+} as a counterion and staining at pH 5.8, the type of polyanion in a tissue may be assessed (Quintarelli and Dellovo, 1965). The epitheliomesenchymal junction of whole salivary glands stained blue with Alcian Blue 8 GX at $MgCl_2$ concentrations of 0.2, 0.3, and 0.4 M (Fig. 2b–d). At concentrations of 0.6 and 0.7 M (Fig. 2e, f), the blue staining of the junction is markedly less intense and the small amount of stain remaining does not completely encircle the epithelium. At these Mg^{2+} concentrations, some sections reveal no staining at the epithelial surface. At still higher levels of Mg^{2+} (0.9 M) no Alcian blue staining occurs at the interface.

The studies of Scott, Quintarelli, and co-workers (see Quintarelli, 1968) on the electrolyte concentrations required to reverse the dye–polyanion complex in tissues demonstrate that loss of stain at 0.2 M $MgCl_2$ generally indicates that Alcian Blue was bound to a polyanion of low charge density or to a weakly ionized polymer, such as a carboxylated polyanion. Polyphosphates (e.g., polynucleotides) destain at slightly higher salt concentrations, and sulfated polyanions lose their stain at even greater concentrations; i.e., the sulfated acid mucopolysaccharide–dye complex is a stronger one. Comparison of the salivary gland staining with studies on other tissues (Quintarelli and Dellovo, 1965) strongly suggests that the epitheliomesenchymal junction contains a sulfated, or mixed, carboxylated and sulfated polymer.

Two-step PAS. The presence of polymeric uronic acid at the epitheliomesenchymal junction was investigated by the use of a method based upon differential oxidation with periodic acid which distinguishes uronic acid-containing compounds from neutral polysaccharides and glycoproteins (Scott and Dorling, 1969). Neutral sugars are rapidly oxidized by periodate to Schiff-reactive aldehydes, while 1:4 linked uronic acid residues in polymers are slowly and specifically converted to Schiff-positive compounds by oxidation of the C_2–C_3 glycol group of the uronic acid (Scott and Harbinson, 1969). This difference in rate of oxidation, coupled with reduction of the rapidly oxidized residues to primary alcohols (which are stable to further periodate oxidation), provides the basis for the differential staining method. Tissues can be oxidized with periodate for a short time, and the aldehydes produced can be made Schiff-negative by reduction with sodium borohydride. A longer periodate oxidation follows, producing Schiff-positive compounds from the slowly reactive uronic acid residues in acid mucopolysaccharides.

This two-step PAS method was used to evaluate salivary glands for acid mucopolysaccharide. The duration of the initial periodate treatment was varied (15–60 minutes) to ensure complete oxidation of the rapidly reactive glycol residues. An initial oxidation of 15 minutes was sufficient, as no further Schiff-reactive material was demonstrable after borohydride reduction and a second 60-minute oxidation (Fig. 3). However, a second periodate treatment of 24 hours yielded definite Schiff reactivity at the epitheliomesenchymal interface. The distribution of this Schiff-positive material was identical with that of Alcian Blue staining at 0.4 M $MgCl_2$, strongly

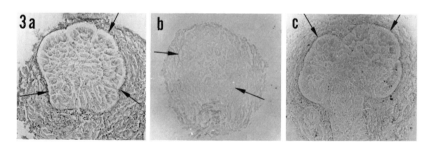

FIG. 3. Staining of the salivary glands by the two-step periodic acid–Schiff method. × 49. (a) Schiff reagent staining of the epitheliomesenchymal junction (arrows) after a 15-minute periodate oxidation. (b) Schiff reagent staining of a section oxidized as in a, then treated with sodium borohydride and reoxidized for 60 minutes. Neither the interface (arrows) nor any other area reacts with the Schiff reagent. (c) Schiff reagent staining of a section oxidized as in a, then reduced with sodium borohydride as in b and reoxidized for 24 hours. The interface (arrows) is stained as in the sections treated with Alcian Blue at $MgCl_2$ concentrations below 0.6 M (compare Fig. 2).

suggesting the presence of acid mucopolysaccharide at the epitheliomesenchymal junction.

Labeling of the Epitheliomesenchymal Junction

The incorporation of glucosamine-^{3}H and $^{35}SO_4$ was studied to determine whether the material staining with Alcian Blue and double-PAS was labeled by these mucopolysaccharide precursors. Whole glands were incubated for a brief period in the presence of glucosamine-^{3}H or $^{35}SO_4$. Rudiments labeled without prior culturing showed a heavy incorporation of glucosamine-^{3}H at the epitheliomesenchymal interface, but little incorporation of $^{35}SO_4$. Rudiments labeled after 1 day in culture clearly demonstrated both radiosulfate and glucosamine-^{3}H radioactivity at the epithelial surface. The distribution of both glucosamine-^{3}H and $^{35}SO_4$ was identical; the greatest localization of radioactivity was at the junction of the developing epithelium and mesenchyme.

Glands labeled for 2 hours immediately after dissection were analyzed to determine whether the rate of glucosamine-^{3}H incorporation differed at various sites of the epitheliomesenchymal junction. Maximal incorporation of label was at the epithelial surface of the distal ends of the growing and branching lobules (Fig. 4a). Incorporation was also noted within the lobules, at the interlobular areas, and on the stalk, but very little label was seen in the mesenchyme. Treatment of the sections with testicular hyaluronidase removed the label primarily from the epithelial surface and the stalk (Fig. 4b). Thus, the radioactivity within the lobule is not in hyaluronidase-susceptible materials. Identical results were obtained by treating the sections with the bacterial hexosaminidases, chondroitinase AC, and chondroitinase ABC (Yamagata *et al.*, 1968). It is of interest that there is substantially greater glucosamine incorporation into hexosaminidase-sensitive material at the surface of the morphogenetically active areas. This difference between sites in the amount of newly synthesized mucopolysaccharide contrasts with the apparently equivalent distribution of mucopolysaccharide as revealed by staining. The sites of most rapid mucopolysaccharide accumulation are those that show the least amount of collagen (see above).

Evidence for acid mucopolysaccharide at the epithelial surface is provided by specific staining for polymeric sulfate and uronic acid, by localization of polysaccharide precursors with autoradiography, and by susceptibility of the incorporated label to three distinct

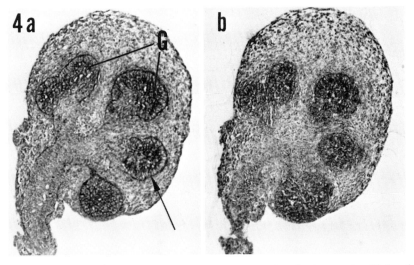

FIG. 4. Autoradiographs of salivary glands incubated with glucosamine-^{3}H for 2 hours. × 120. (a) Substantial amounts of radioactivity are within the lobule (G) and at the epitheliomesenchymal junction of the distal ends of the lobules (arrow). Considerably less label is at the surface of the interlobular clefts and of the stalk. There is minimal incorporation into the mesenchyme. (b) A section as in a, treated with testicular hyaluronidase. The label is removed from the epithelial surface, but not from within the lobules. From Bernfield and Banerjee (1971a).

endo-β-hexosaminidases. Moreover, material with chemical characteristics[1] identical with those of acid mucopolysaccharide could be isolated from the glands.

SURFACE-ASSOCIATED MUCOPOLYSACCHARIDE AND SALIVARY GLAND MORPHOGENESIS

Epithelia free of mesenchyme were isolated from intact salivary glands by repeated flushing during treatment with crude trypsin-pancreatin, crystalline trypsin, and purified collagenase. Treatment of glands with EDTA was not suitable because concentrations of EDTA that removed mesenchymal cells also removed cells from the epithelium. Hyaluronidase treatment over a wide concentration range did not remove the investing mesenchyme. The isolated epi-

[1] The material was labeled with both glucosamine-^{3}H and $^{35}SO_4$, and was isolated by procedures involving RNase and DNase, as well as pronase and papain digestions. It was not extractable with acetone-ether, and the precipitate which formed with cetyl pyridinium chloride was methanol soluble in the absence of salt.

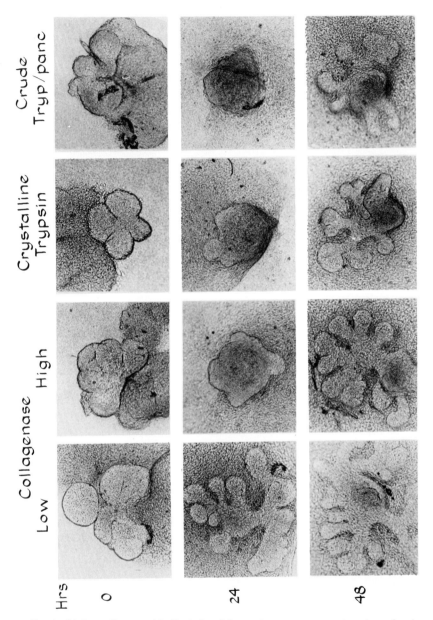

FIG. 5. Living salivary epithelia isolated by various enzyme treatments and cultured in direct combination with salivary mesenchyme. Epithelia isolated with low concentrations of collagenase continue morphogenesis unabated. Epithelia isolated with high concentrations of collagenase, crystalline trypsin, or crude trypsin-pancrea-

thelia were cultured without a plasma clot either on top of mesenchyme pieces or directly on Millipore filters in the absence of mesenchyme. Intact, but enzyme-treated glands were cultured directly on the filter (Bernfield and Banerjee, 1971b). During culture in direct combination with mesenchyme, epithelia isolated by trypsin-pancreatin treatment (0.3–3.0%) lost their characteristic shape (Fig. 5). The clefts between developing lobules disappeared, and within 16–20 hours the epithelia formed a ball-like structure. With continued culture, the rounded epithelial mass produced budding outgrowths which underwent progressive branching. The distal ends of these branches produced bulbous adenomeres which subsequently developed acini. Failure to maintain the original morphology or to continue morphogenesis unabated was also observed with epithelia prepared with crystalline trypsin (0.001–0.05%) (Fig. 5).

The morphogenetic behavior of epithelia isolated by collagenase treatment varied with both the enzyme preparation and the enzyme concentration. To clarify this variability, each of five collagenase preparations was assessed at several enzyme concentrations. For three of the preparations (1SS, 9EA, and 9BB), collagenase concentrations were found that yielded epithelia which maintained their shape, retained the clefts between lobules, and underwent uninterrupted morphogenesis when cultured with mesenchyme (Fig. 5). Epithelia isolated with higher concentrations of these preparations, or with any concentration tested of preparations 9BA or 9EB (0.005–0.07%) rapidly lost their characteristic contour and formed a spherical mass of tissue (Fig. 5). This occurred despite the continuous presence of salivary mesenchyme. At about 20–24 hours of culture, the ball-like epithelial rudiments began to produce outgrowths which subsequently branched in a characteristic fashion and yielded adenomeres. The epithelia prepared with high concentrations of collagenase lost their contours more rapidly than those described above that were separated from mesenchyme with trypsin-pancreatin or crystalline trypsin. Some high collagenase-isolated epithelia were observed to change their shape within several minutes after removal of mesenchyme, although generally the initial effects were seen in 4–6 hours.

The epithelia isolated with low collagenase concentrations dif-

---

tin lose their lobules and form "ball-like" rudiments, despite the continued presence of mesenchyme which surrounds the epithelia. At 24 hours outgrowths have formed from these epithelia, and at 48 hours these outgrowths have formed branches. × 33. From Bernfield and Banerjee (1971b).

fered from all other isolated epithelia in continuing characteristic salivary morphogenesis when recombined in culture with fresh mesenchyme. Intact glands subjected to identical enzyme treatments, but not flushed to remove mesenchyme, all developed normally. Thus, this difference was apparently not due to lower nonspecific toxicity of the low collagenase treatment. Careful histologic observations did not reveal any differences between epithelia isolated with the various enzymes in the number of adhering mesenchymal cells. Indeed, electron microscopy of epithelia isolated by low collagenase treatment showed essentially no mesenchymal cells at the epithelial surface (see below).

Although these results are similar to those reported by Grobstein and Cohen (1965) and Wessells and Cohen (1968), some significant differences should be noted. These workers used slightly younger salivary glands where multiple lobes were not evident at time of culture. Epithelia were freed of mesenchyme by trypsin-pancreatin treatment (3.0%) and were cultured in a plasma clot *transfilter*) from mesenchyme. Exposure to collagenase was either continuous, by inclusion of collagenase in the medium, or for short periods, usually 48 hours after the cultures had been established. Under these conditions, collagenase treatment yielded a flat, sheetlike, smoothly contoured epithelium. The effect of cysteine, an irreversible inhibitor of collagenase (Seifter *et al.*, 1959), in the medium cannot be evaluated because the cysteine concentration of medium containing chick embryo extract is unknown.

The present results contrast with these in that removal of mesenchyme with trypsin-pancreatin (0.3%), crystalline trypsin, or high collagenase concentrations yielded epithelia which formed a spherical, ball-like mass during culture in direct association with fresh salivary mesenchyme. Epithelia isolated with low concentrations of collagenase maintained their characteristic shape and underwent continued morphogenesis.

Mucopolysaccharide at the Surface of Isolated Epithelia

The morphogenetic behavior of the low collagenase-isolated epithelia in culture combined with mesenchyme could be due to the retention of extracellular substances, while the loss of morphogenesis of trypsin-pancreatin-, crystalline trypsin-, and high collagenase-isolated rudiments could be due to the removal of this material. Since treatment with collagenase (Wessells and Cohen, 1968) and trypsin-pancreatin (Kallman and Grobstein, 1966) removes the epithelial basement membrane and associated materials, and since acid mucopolysaccharide was localized at the epithelial surface, the possibility that surface-associated acid mucopolysaccharide is required for maintenance of morphology and for continued morphogenesis was investigated (Bernfield and Banerjee, 1971b).

The presence of mucopolysaccharide at the surface of epithelia which lost their shape was compared with that of epithelia which retained their surface contour and continued to develop. Intact rudiments were incubated with glucosamine-^{3}H prior to removing the mesenchyme and autoradiography was performed after the various enzyme treatments. Significant radioactivity was observed adjacent to the surface of low collagenase-isolated epithelia (Fig. 6). Maximal localization of the label was at the surface of the growing lobules, as had previously been observed in intact glands. The trypsin-pancreatin- and high collagenase-isolated epithelia were nearly free of surface-associated glucosamine radioactivity. Hyaluronidase treatment of sections of low collagenase-isolated epithelia removed the corona of radioactivity without substantially altering the amount of label within the lobules, demonstrating that this

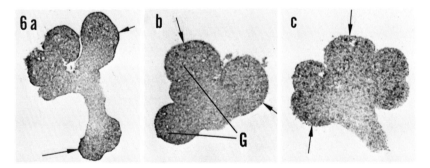

FIG. 6. Autoradiographs of salivary epithelia incubated as intact glands in glucosamine-^{3}H and then isolated free of mesenchyme by various enzyme treatments. × 97. (a) Epithelium isolated by treatment with *low* concentrations of collagenase. Radioactivity is localized at the surface of the lobules (arrows) and within the lobules in a distribution essentially identical with that shown in Fig. 4a. (b) Epithelium isolated by treatment with *high* concentrations of collagenase. Although label is seen within the epithelium (G), the surface of the lobules (arrows) is nearly free of radioactivity. (c) Epithelium isolated by treatment with crude trypsin-pancreatin. The epithelial surface (arrows) is essentially devoid of radioactivity, while label remains within the lobules.

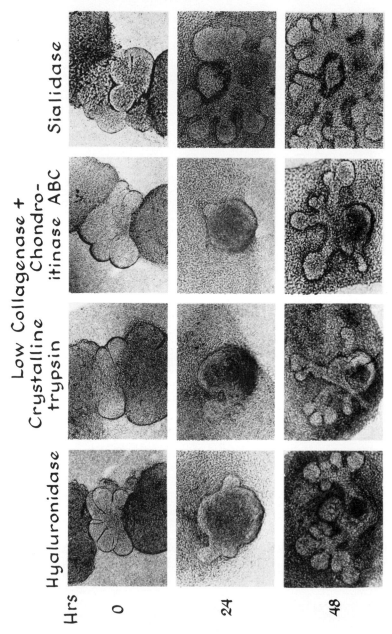

FIG. 7. Living salivary epithelia isolated with low concentrations of collagenase and exposed to various enzymes prior to culturing in direct combination with salivary mesenchyme. Epithelia exposed to hyaluronidase, crystalline trypsin, and chondroitinase ABC lose their lobules and form "ball-like" epithelia which at 24 hours have produced outgrowths. At 48 hours, the outgrowths have branched. This sequence is very

material behaves identically to that identified as acid mucopolysaccharide in intact glands.

Treatment of Low Collagenase-Isolated Epithelia with Various Enzymes

To further assess the nature of the surface-associated material and its relationship to morphogenesis, epithelia freed of mesenchyme by low collagenase treatment were exposed (without flushing) to crystalline trypsin, hyaluronidase, sialidase, and two bacterial enzymes capable of degrading mucopolysaccharides (Yamagata et al., 1968), chondroitinase AC, and chondroitinase ABC. The glands were then combined with fresh mesenchyme for culture. Of these treatments, only the control epithelia and those treated with sialidase showed continued morphogenesis (Fig. 7). Doubling the concentration of sialidase gave identical results. In none of these treatments did the second enzyme either fragment the epithelium or prove toxic as determined by darkening of the tissue.

The time sequence of events during the loss of epithelial structure and resumption of morphogenesis depended on the type of enzyme treatment. Hyaluronidase and chondroitinase treatments caused a fairly rapid loss of epithelial contour. Depending on the degree of lobulation of the epithelium, loss of shape could be noted within 1–2 hours after hyaluronidase treatment. These epithelia began to show outgrowths from the spherical mass at approximately 12–16 hours of culture. However, the outgrowths did not branch until about 20–26 hours. The hyaluronidase-treated epithelia behaved differently in this respect from other epithelia prepared by two-stage treatment, or from epithelia freed of mesenchyme with trypsin, trypsin-pancreatin or high concentrations of collagenase. The ball-like epithelia developing from these latter rudiments formed outgrowths more slowly, but branching of the outgrowths followed without delay.

Mucopolysaccharide at the Surface of Isolated Epithelia Treated with Various Enzymes

The two-stage enzyme-treated epithelia were assessed for surface-associated material in an identical fashion as before. Intact glands

---

similar to that depicted in Fig. 5. Epithelia exposed to sialidase do not lose their characteristic shape; they continue to undergo morphogenesis, in an identical fashion as low-collagenase treated epithelia not exposed to a second enzyme treatment. × 40. From Bernfield and Banerjee (1971b).

incubated with glucosamine-³H were freed of mesenchyme with low concentrations of collagenase, and the isolated epithelia were then treated with the second enzyme and fixed for autoradiography. Glucosamine-³H radioactivity was observed at the surface of the control epithelia and surrounding the epithelia treated with sialidase. However, very little radioactivity was localized at the surface of the hyaluronidase-treated epithelia (Fig. 8). Thus, there is a strong reciprocal relationship between the presence of surface-associated mucopolysaccharide, as indicated by autoradiography, and the maintenance of epithelial shape as well as the continuation of uninterrupted morphogenesis.

Microscopy of Epithelia

The histologic and ultrastructural characteristics of the isolated and cultured epithelia subjected to various enzyme treatments were examined. Light microscopy revealed no significant differences between the epithelia fixed immediately after low-collagenase or low-collagenase plus hyaluronidase treatment. After 17 hours of culture combined with salivary mesenchyme, the low collagenase-treated epithelia showed distinct lobules. There was a nearly contiguous layer of columnar cells at the basal surface of the epithelium (Fig. 9a). Numerous mitotic figures were seen in the cells subjacent to the basal cell layer. The low collagenase plus hyaluronidase-treated rudiments at 17 hours of culture were rounded, amorphous masses of disorganized cells (Fig. 9b). There was little evidence of a discrete basal columnar cell layer, and the cells appeared mixed to-

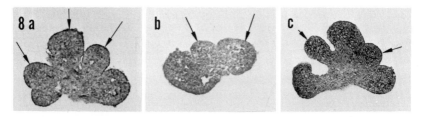

FIG. 8. Autoradiographs of salivary epithelia incubated as intact glands in glucosamine-³H, isolated free of mesenchyme with a low concentration of collagenase and then exposed to various enzymes. × 66. (a) Isolated epithelium exposed to Ca-Mg-free Tyrode's solution, the solvent for the enzymes described below. Radioactivity is localized at the epithelial surface (arrows) as in Figs. 4a and 6a. (b) Isolated epithelium exposed to hyaluronidase. The surface (arrows) is nearly free of label. (c) Isolated epithelium exposed to sialidase. Radioactivity is at the surface (arrows) of the lobules.

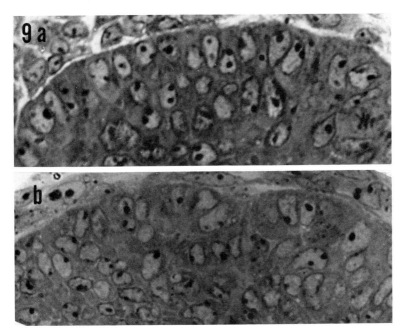

FIG. 9. Views near the epitheliomesenchymal junction of isolated epithelia *cultured for 17 hours* in direct combination with mesenchyme. × 560. (a) Epithelium isolated with a low concentration of collagenase. There is a layer of columnar cells at the basal surface. A mitotic figure is seen just below this layer (lower right). (b) Epithelium isolated as in a but exposed to hyaluronidase prior to culture. The basal columnar cell layer is disorganized. Very few mitoses are seen in such epithelia.

gether, rather than ordered, as in the low-collagenase isolated epithelia. There were substantially fewer mitotic figures. The loss of cellular orientation in these epithelia is reminiscent of the substratum requirement for basal epidermal cells to remain oriented and mitotically active (Wessells, 1964). The situation observed here may be analogous to the observation (Slavkin et al., 1969b) that disaggregated cells which contact an extracellular matrix become oriented toward the surface of the matrix to form a single layer of columnar cells.

Electron microscopic observations confirmed these results, and demonstrated that the junctional complexes of the lateral surfaces of the cells were intact in both types of rudiments either immediately after enzyme treatment or after 17 hours of culture. The 40–50 Å microfilaments that will be described in detail below, insert into

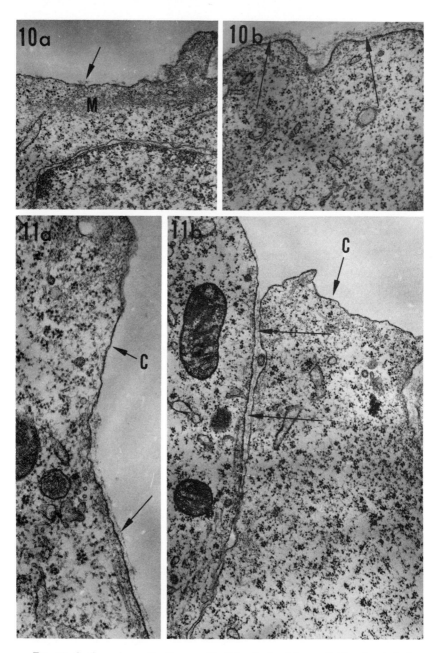

FIG. 10. Surface views of salivary epithelial cells fixed immediately after isolation with a low concentration of collagenase. Most of the epithelial surface is as shown in b. However, areas apparently lacking most extracellular materials (as in a) are also found. Note the microfilament bundle in a. × 25,000.

these junctional complex regions, and remain organized in bundles. Epithelia fixed for electron microscopy immediately after enzyme treatment revealed a clear difference in surface-associated material between the low collagenase and low collagenase plus hyaluronidase-treated rudiments (Figs. 10 and 11).

The *low collagenase-treated* epithelia (Fig. 10) showed a distinct electron-dense basal laminar layer surrounding much of the epithelial surface. This electron-dense material averages 400 Å in thickness and is located some 300 Å from the plasma membrane of the epithelial cells. The density of the basal lamina is definitely less than in control salivary glands, indicating that some materials were removed by low collagenase treatment. The extracellular materials normally found peripheral to the basal lamina of untreated glands are depleted by low collagenase treatment so that only patches of these substances are found scattered over the tissue surface. In a few places, low collagenase treatment removes all visible extracellular materials from the tissue.

The basal lamina and adjacent materials were absent from most surface areas of the *low collagenase plus hyaluronidase-treated* epithelia (Fig. 11). However, a few areas were found which retained surface-associated extracellular materials. Despite the difference in cell organization and orientation between the 17-hour cultured epithelia, substantial quantities of extracellular material and a nearly complete basal lamina were seen in both epithelial types after 17 hours of culture (Fig. 12 and 13). Again, less extracellular material was seen in these epithelia than in glands from which mesenchyme had not been removed.

These observations demonstrate that disruption and loss of the basal lamina and associated extracellular materials correlates with subsequent loss of morphology and cellular orientation. Retention of these components is associated with maintenance of epithelial morphology and continued morphogenesis. Furthermore, it should be noted that reappearance of these surface-associated components in the low-collagenase plus hyaluronidase-treated epithelia apparently preceded the reorganization of cells and the initiation of

---

FIG. 11. Typical surface regions after isolation with a low concentration of collagenase followed by exposure to hyaluronidase. Most areas (C) are devoid of surface-associated substances. Remnants of such materials occur frequently in clefts and occasionally on the surface of the tissue (arrows). Even when present, such material appears less electron dense than in similar epithelia not exposed to hyaluronidase. a, × 33,000 b, × 22,000.

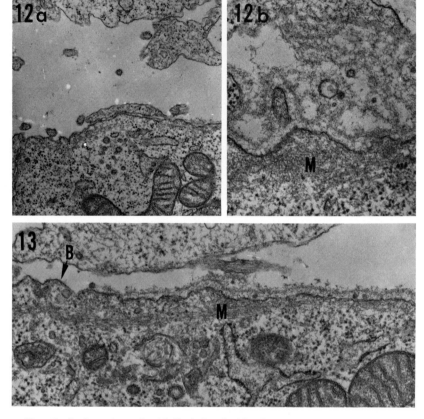

Fig. 12. Surface views of epithelia isolated with a low concentration of collagenase, and cultured for 17 hours *in vitro* in the presence of mesenchyme. These two views show typical variability in quantity of extracellular materials seen on such epithelia. Microfilaments (M) are seen in a, $\times$ 21,000; b, $\times$ 36,000.

Fig. 13. Surface view of an epithelium isolated with low-collagenase, exposed to hyaluronidase, and then cultured for 17 hours *in vitro* in the presence of mesenchyme. Variable quantities of extracellular materials have accumulated during culture (compare with Fig. 11a, b), but scattered "bare" areas (B) are still found. Microfilaments (M) course beneath the plasma membrane of this cell. $\times$ 38,000.

the branching process, suggesting that reconstitution of the basal lamina is a prerequisite for the resumption of characteristic morphogenesis.

Degradative Activities of Enzymes

Several different enzymatic treatments were used in establishing the correlation between surface-associated mucopolysaccharide and the maintenance of epithelial morphology. The collagenolytic, proteolytic, and polysaccharidolytic activities of these enzymes bear heavily on the interpretation of the experiments (Bernfield and Banerjee, 1971b). Consequently, each of the enzymes was assayed for protease activity (denatured hemoglobin substrate, Davis and Smith, 1955), tryptic esterase activity (p-tosyl-L-arginine methyl ester, Hummel, 1959), mucopolysaccharidase activity (crude chondroitin sulfate), and collagenase activity (4-phenylazobenzyloxycarbonyl-L-leucyl-glycyl-L-prolyl-D-arginine, Wünsch and Heidrich, 1963). Mucopolysaccharidase activity was measured by a turbidometric method (Mathews, 1966) as well as for bacterial enzymes, by a β-elimination method (Linker, 1966).

The clostridial collagenase preparations demonstrated insignificant tryptic esterase activities, confirming the results of Grobstein and Cohen (1965) and Wessells and Cohen (1968) (Table 1). However, noncollagen proteolytic activity was detected in all commercial preparations (9EA, 9BB, and 9BA). The collagenase preparation devoid of protease activity (1SS) was an enzyme prepared by a modification of the procedure of Harper et al. (1965). This enzyme

TABLE 1
Degradative Activities of Enzymes[a]

Enzyme	Protease $\left(\dfrac{\text{nmole}}{\text{min mg}}\right)$	Esterase $\left(\dfrac{\Delta A_{247}}{\text{min mg}}\right)$	Mucopolysaccharidase		Collagenase $\left(\dfrac{\mu\text{mole}}{\text{min mg}}\right)$
			Turbidometric $\left(\dfrac{\Delta A_{600}}{\text{min mg}} \times 10^3\right)$	β-Elimination $\left(\dfrac{\Delta A_{235}}{\text{min mg}} \times 10^2\right)$	
Collagenase: 1SS	0.0	—	41.6	28.5	2.2
9EA	16.3	<0.001	12.9	9.41	1.4
9BB	11.7	<0.001	14.7	7.98	3.3
9BA	75.7	<0.001	43.6	29.4	3.2
Crystalline trypsin	610	150	0.0	—	0.0
Crude trypsin-pancreatin	108	0.67	106	—	0.0022
Hyaluronidase	0.0	—	194	—	0.0
Chondroitinase ABC[b]	0.0	—	53.0	32.1	0.0
Chondroitinase AC[b]	0.0	—	125	85.7	<0.001
Clostridial sialidase	14.9	—	0.0	0.0	0.0

[a] Protease, tryptic esterase, mucopolysaccharidase and collagenase assays of the enzymes used for isolation of epithelia and for treatment of isolated epithelia. Assays were performed as cited in the text.

[b] Activities for chondroitinase ABC and AC are calculated per unit enzyme (Yamagata et al; 1968).

was apparently monodispersed by polyacrylamide gel electrophoresis, sedimentation equilibrium ultracentrifugation, and agar gel immunodiffusion (Lent and Seifter, 1969).

All the collagenase preparations demonstrated appreciable mucopolysaccharidase activity, whether measured by degradation (turbidometric assay) or formation of unsaturated disaccharides (β-elimination assay). The mucopolysaccharidase activity was not proportional to the collagenolytic activity of the preparations. Crystalline trypsin was devoid of mucopolysaccharidase or collagenase activity. Crude trypsin-pancreatin, in addition to tryptic activity, showed considerable mucopolysaccharidase activity. The preparations of testicular hyaluronidase, *Proteus vulgaris* chondroitinase ABC, and *Flavobacterium heparinum* chondroitinase AC were free of protease and collagenase activity. The *Clostridium perfringens* sialidase preparation demonstrated some protease activity, but was free of mucopolysaccharidase or collagenase activity. These assays show that clostridial collagenase preparations contain substantial mucopolysaccharidase activity. However, there is no apparent relationship between the amount of protease and/or mucopolysaccharidase activities contaminating a collagenase preparation and the effect of various concentrations of that preparation upon morphogenesis. Possibly, this is because the precise nature of the morphogenetically active material and the mechanism by which it is removed are not known. Furthermore, it is likely that there is more than a single substance involved, and that the morphogenetic behavior of the epithelia reflects a combination of factors, some of which may require a highly ordered assembly of macromolecules. Nevertheless, it is clear that *the dependence of morphogenesis on "collagenase-susceptible" materials does not necessarily implicate collagen.*

CHARACTERISTICS OF THE SURFACE-ASSOCIATED MATERIAL

What are some properties of the extracellular morphogenetic matrix? The enzymes which result in loss of morphology when used in the second step of the two-stage assay are enzymes that degrade proteins and mucopolysaccharides. The specificity of these enzyme treatments and the demonstration that the presence of surface-associated acid mucopolysaccharide correlates with epithelial morphogenesis strongly suggests that morphogenesis is dependent on acid mucopolysaccharide-protein complexes. Trypsin treatment of cells is known to solubilize varying amounts of sialic acid (Kraemer,

1966), and this compound is a component of membrane glycoproteins (Cook, 1968) and the cell surface coat (Wu et al., 1969). Thus, lack of a morphogenetic effect of sialidase supports the data implicating a mucopolysaccharide–protein complex. Because of the possibility that a contaminating ribonuclease (RNase) activity might be exerting a morphogenetic effect (Weiss and Mayhew, 1966; Mayhew and Weiss, 1968; Slavkin et al., 1969a), the enzyme preparations were examined for RNase activity by an ultrasensitive procedure (Zimmerman and Sandeen, 1965).

Crystalline trypsin demonstrated minimal RNase activity, but the hyaluronidase, chondroitinase ABC and chondroitinase AC preparations were completely devoid of RNase activity. It is possible that another contaminating activity (e.g., glycosidase, sulfatase, lipase) is responsible for the cessation of morphogenesis. This explanation cannot be rigorously excluded, but is unlikely since the polysaccharidase preparations were derived from three distinct sources by different procedures, and both bacterial enzymes were purified to apparent homogeneity (Yamagata et al., 1968).

With the exception of hyaluronic acid, animal acid mucopolysaccharides are covalently linked to protein (Thorp and Dorfman, 1967; Rodén, 1968). Thus, proteolytic treatment might result in loss of mucopolysaccharide from the epithelial surface. It is premature to suggest which component of a mucopolysaccharide–protein complex is responsible for maintenance of morphology, or which of the known acid mucopolysaccharides is involved. This is particularly true, since the acid mucopolysaccharides demonstrate a considerable degree of molecular heterogeneity (Rodén, 1968).

The presence of a mucopolysaccharide-protein conjugate at the epithelial surface supports the hypothesis of previous workers that epithelial collagen fibrogenesis is due to some property of the epithelial surface (Edds and Sweeny, 1961; Grobstein, 1967; Bernfield, 1970). In vitro interactions of collagen with acid mucopolysaccharide (Mathews, 1965) as well as with mucopolysaccharide-protein complexes (Toole and Lowther, 1968a) have been described. The latter interaction results in the formation of "native-type" collagen fibrils which form coarse bundles (Toole and Lowther, 1968b). It is impossible, at this time, to assign levels of importance to various components of the extracellular matrix. However, in transfilter cultures, collagen fibers are seen only infrequently, whereas basement membrane and associated materials cover almost the entirety of the epithelial surface (Wessells and Cohen, 1968). Collagen fibers are

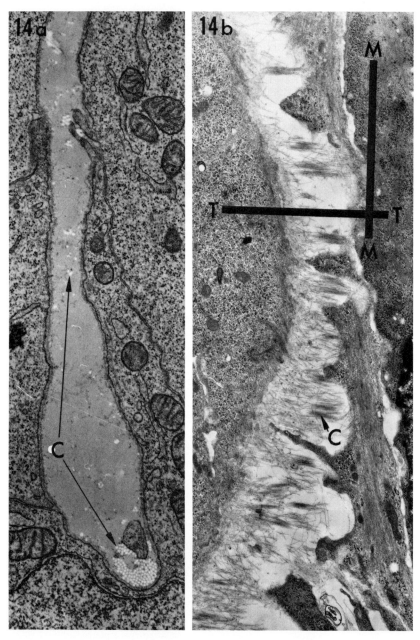

FIG. 14. Examples of collagen distribution during epithelial morphogenesis. (a) In the deep cleft in the surface of a salivary epithelium, a large bundle of collagen is cut in cross section at the very base of the furrow. Similar tightly packed bundles are exceedingly rare in any other portion of the developing salivary gland. (b) A section cut

more abundant adjacent to morphogenetically quiescent areas of the lung (Wessells, 1970) and salivary epithelium (Grobstein and Cohen, 1965), while newly synthesized acid mucopolysaccharide predominates at the surface of morphogenetically active zones (Figs. 4 and 14). The features of the epithelial surface responsible for these distributions are wholly unknown. The characteristics of collagen and mucopolysaccharide distribution may be significant in understanding the mechanism of morphogenesis, as is discussed below.

INTRACELLULAR CONTROL OF MORPHOGENESIS

Two kinds of intracellular organelles are prominent candidates for the agents that control cell shape during morphogenesis. Microtubules appear to play an essential role in the maintenance of asymmetric cell shape, and perhaps in generating changes in cell shape, as during lens placode formation (Byers and Porter, 1964), or spermatid elongation (McIntosh and Porter, 1967). A clear role for microtubules has not been discerned in the complex three-dimensional changes in shape of cell populations as seen during development of salivary gland, lung, or kidney, or in analogous stages of lens morphogenesis (Pearce and Zwaan, 1970). In epithelia of these organs, microtubules are involved in spindle function and in maintaining the asymmetric columnar shape of the epithelial cells.

Microfilaments are the other class of intracellular organelles that may actively participate in morphogenesis. These filaments average 40–50 Å in diameter and occur in bundles at discrete places within developing cells. In columnar cells of embryonic endoderm, for instance, they are abundant at both the luminal and basal ends of cells. They appear to insert into amorphous, electron-dense material located just internal to the plasma membrane at the luminal end at the level of the zonula adhaerens, and at the basal end in an equivalent area distal to the desmosomal zone. Microfilaments differ from the filaments that loop (Kelly, 1966) through the plaques of desmosomes in size (ca. 50 Å versus 100 Å in diameter), in sensitivity to cytochalasin B (see below), and in general distribution within cells (bundles of the 100 Å filaments may course in a variety of directions

tangential to the surface of developing trachea. The long axis of the trachea runs from left to right (T), and the predominant orientation of collagen (C) parallels the tracheal axis. The layer of mesenchymal cells next to the trachea are highly elongated, and their long axes run perpendicular (M) to that of the collagen fibers. These distributions are seen in all "morphogenetically quiescent" regions of the tracheobronchial tree (see Wessells, 1970). a, × 19,500; b × 7500.

through central cytoplasmic regions). In most preparations, microfilaments appear to be short, straight structures that are of indeterminate length; it is uncertain whether this is due to their passage in and out of the plane of section or to actual short length.

Microfilaments are thought to be active agents of morphogenesis because of their putative contractile properties. Cloney's spectacular movie and publication (1966) on the retraction of tadpole tails during metamorphosis of the ascidian, *Amaroucium*, provide the strongest case for such a contractile function. The tadpole epidermal cells decrease from ca. 26 μ to ca. 1–2 μ in length in 6 minutes, and, as the cells change shape, a thickening band of microfilaments appears across their outer ends. Microtubules, thick cytoplasmic filaments, and other organelles show no structural modifications that correlate with the change in cell shape. Interestingly, tails isolated from the tadpole bodies can still display the retraction phenomenon: the epidermal tissue appears to contract, forcing the tail contents out into seawater, in the same way that the contents are normally propelled into a cavity in the metamorphosing tadpole's body. Recently, Cloney (1969) made the discovery that epidermal cells of *Boltenia villosa* do not contract during metamorphosis and those cells lack a microfilament assembly. In *Boltenia*, it is notochordal cells that change in shape very rapidly, and they contain skeins of microfilaments that widen concomitant with the alteration in cell shape. In both species, the alterations in cell and tissue morphology can be explained most simply by the hypothesis that the filaments are contractile. The microfilaments, for instance, are located in the precise places where contractile activity could account for the changes in cellular dimensions.

Further data favoring the idea that microfilaments are contractile and are involved in changing cell shape come from ultrastructural studies of a number of cell types participating in morphogenetic movements.

Barker and Schroeder (1967), in a landmark paper, demonstrated that contraction of the 50 Å microfilaments would result in rolling up of the flat medullary plate of amphibian embryos into the neural tube. An analogous movement is thought to occur as the flat lens placode invaginates into a lens cup (Wrenn and Wessells, 1969). In both these cases, skeins of microfilaments surround the upper, outer ends of the cells.

Contraction of these filaments, in the manner of a "pursestring" (Baker and Schroeder, 1967), would decrease the cross-sectional area of the outer end of individual cells. Since the lateral surfaces of the cells are apparently attached firmly to each other in that region by some form of junctional complex, the result of microfilament contractility would be an imposition of curvature upon the formerly flat surface.

In these studies and in virtually every other example in which the filaments seem to function in an analogous way, the surface of the cells near the filament bundles is twisted and convoluted. In addition to neural tube and lens cup formation, such surface convolutions have been observed during amphibian gastrulation (Baker, 1965), salivary gland morphogenesis (Spooner and Wessells, 1970a), tubular gland formation in oviduct (Wrenn and Wessells, 1970), and in tadpole tail shortening (Cloney, 1966, 1969). The cell surface convolutions are probably a transient phenomenon resulting from a sufficiently rapid contraction of the microfilaments that compensatory reduction in cell surface area to keep the surface smooth and flat is not observed. This phenomenon will be discussed further below.

The most important implication from observations such as Baker and Schroeder's is that contraction of microfilaments can be selective in terms of space—that is, in discrete populations of cells—and in terms of time—that is, within a circumscribed developmental period. The result of these discretely controlled events is a multicellular phenomenon that is recognized as a morphogenetic movement.

Microfilaments at the luminal ends of cells are also thought to function in the alteration in cell shape that accompanies narrowing of the initial pancreatic duct, a process that can rightfully be considered the initial morphogenetic movement of pancreas formation (Wessells and Evans, 1968a). Interestingly, equivalent sets of filaments are seen in other cells of the pancreatic anlage, *not* engaged in changing shape, as well as in nearby cells of the prospective stomach and intestine. Bundles of filaments are found in all epithelial cells of the tracheobronchial tree during lung formation (Wessells, 1970). Experiments with the drug cytochalasin (see below) have revealed that microfilaments are a cellular organelle common to most if not all cells, and are capable of being used for a variety of cellular or developmental processes. Thus, during initial formation of the pancreas, it may be assumed that the contractile microfilament assembly alters the shape of only a restricted population of cells, thereby narrowing the pancreatic duct, but leaving the bulbular portion in an expanded condition.

Obviously, the factors that lead to synthesis, assembly, and function of the microfilament arrays are of key importance in understanding the role of these organelles, and, more important, the cellular basis of morphogenesis. Experiments by Wrenn and Wessells (1970) provide the first case in which filaments appear in response to a known agent. In oviducal epithelium of young chicks (3–10 days posthatching), 50 Å microfilaments are not seen near the junctional complex region at the luminal end of the columnar epithelial cells. At this time, the oviduct is quite round and smooth and tubular glands, the structures involved in ovalbumin production and secretion, are not evident. Thirty-six hours after a 7-day-old chick is injected with estradiol, aggregates of cells have begun to bulge outward from the oviducal surface into the stromal spaces. A small lumen marks each of these early tubular glands, and secretory granules are invariably seen in the luminal cytoplasm of the cells. Most striking is the presence of thick skeins of microfilaments extending from one side to the other side of the epithelial cells. These filaments are identical in relative position to those seen during neural tube (Baker and Schroeder, 1967) and lens formation (Wrenn and Wessells, 1969). If the filaments are contractile elements and are responsible for the initial evagination of glandular cells, then this observation is of extraordinary interest since it implies that the inducing hormone causes not only the initiation of specific synthesis and differentiation of the cells, but also the assembly of organelles that are responsible for morphogenesis. In general terms, this sytem shows that a determined population of cells has been acted upon by a regulatory agent so that all processes involved in tubular gland formation are initiated. The result is a coordinated process involving both morphogenesis and cytodifferentiation.

Investigations of microfilament function have been extraordinarily expedited as a result of an incisive discovery by Schroeder (1969). Since 1967, it has been known that cytochalasin B causes a variety of cell types to become multinucleate (Carter, 1967; Ridler and Smith, 1968). No satisfactory explanation of the drug's action had been found until Schroeder applied cytochalasin to cleaving marine eggs. Previously, evidence had accumulated that a ring of contractile filaments girdles an egg immediately beneath the cleavage furrow. Szollosi (1970), Arnold (1968), Schroeder (1968), and Rappaport (1967), among others, have provided indirect evidence that these "microfilaments" are the agents responsible for cytokinesis. Schroeder (1969) observed that cytochalasin halts narrowing of the cleavage furrow, thus stopping cytokinesis, and, at the same time, disperses the contractile ring microfilaments. Mitosis itself continues, thus leading to multinuclearity. Presumably the same phenomenon causes the multinuclearity of fibroblasts and lymphocytes exposed to the drug (Carter, 1967; Ridler and Smith, 1968).

MICROFILAMENTS AND SALIVARY GLAND MORPHOGENESIS

We have used cytochalasin to probe the function of other microfilament systems, including ones involved in epithelial morphogenesis (oviduct, salivary gland) and in single cell morphogenesis (nerve). The salivary gland provides a particularly attractive place to investigate the role of intracellular organelles in epithelial morphogenesis, since as outlined above extracellular substances are quite clearly involved in that process. It is not yet known whether the characteristic contour of the salivary epithelium is primarily controlled by such extracellular factors or by the epithelial cells themselves. The following experiments are directed toward an understanding of this problem.

Microfilaments are present in all salivary gland epithelial cells. Thick bundles of the filaments extend across the basal surface of the cells, often just inside the plasma membrane (Fig. 17). The microfilaments insert at the lateral cell surfaces in electron dense material, similar in appearance after lead and uranyl staining, to the material that inserts in the zonula adhaerens portion of the classical junctional complex (Farquhar and Palade, 1963). The filaments are seen in glands undergoing morphogenesis *in vivo*, or in ones explanted in organ culture.

Addition of cytochalasin B to the medium of such organ cultures causes the epithelium, within 6–8 hours, to begin to lose its characteristic shape and to flatten upon the culture substratum. By 18–24 hours this process is complete (Fig. 15), and no further morphogenesis takes place. Two major processes occur during culture in the presence of cytochalasin: (1) the clefts and lobules of the epithelium disappear; (2) the thick, rounded epithelium becomes thin and waferlike. This condition can be seen particularly well after removing such an epithelium from its investing mesoderm with trypsin-pancreatin (Spooner and Wessells, 1970a).

A single ultrastructural alteration occurs within epithelial cells in response to cytochalasin. The bands of microfilaments (Fig. 17) become disorganized, and in their place, at the basal end of the cells,

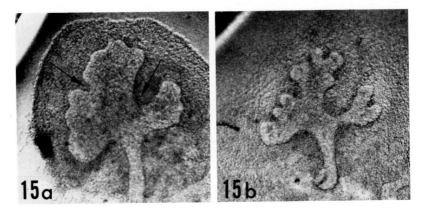

Fig. 15a. A living salivary gland cultured in cytochalasin for ca. 18 hours. This gland was at an advanced stage at the time of explantation, and the deeper surface clefts (arrows) have not completely disappeared (see text). The epithelium is very thin. Note the sharp contours of the mesenchyme, a condition reflecting the absence of normal cell migration from that tissue. (b) The same gland 18 hours after cytochalasin had been removed. Note that a number of new small clefts have appeared along the edges of the epithelium, and that the overall area covered by the epithelium appears smaller than in a; this reflects "rounding-up" of the tissue. The mesenchyme has begun to spread and its edges are now indistinct.

are found large pools or masses composed of short lengths of filaments and possibly granules (Fig. 18). Similar microfilaments at the luminal ends of the same cells are also dispersed into masses of short filamentous material. Microtubules appear normal in these cells, as do thicker (100 Å) filaments associated with desmosomes. Mitochondria, rough endoplasmic reticulum, Golgi apparatus, and other organelles seem unaffected by the drug. Furthermore, cytochalasin produces no demonstrable alteration in rate of amino acid incorporation into hot acid-insoluble materials (Spooner and Wessells, 1970c). Such incorporation is ca. 15% lower than in control cultures, but that reduction is also seen in cultures labeled in the presence of dimethyl sulfoxide, the solvent of cytochalasin (see Carter, 1967). Neither the dimethyl sulfoxide itself nor the slight reduction in protein synthesis it elicits have deleterious effects upon growth or morphogenesis of salivary glands. These results, in combination with others (Yamada et al., 1970; Wrenn and Wessells, 1970) to be discussed below, suggest that altered microfilament function is the critical lesion caused by cytochalasin.

It is noteworthy with respect to the effects of enzymes that remove extracellular materials from salivary epithelium (see above) that neither cytochalasin nor dimethyl sulfoxide alter the morphology of the basal lamina or other extracellular materials as judged by electron microscopy. The flattening of cytochalasin-treated epithelia is not dependent upon gross alteration of collagen or basal lamina, a finding consistent with the hypothesis that the drug acts intracellularly upon microfilaments. Furthermore, as will be discussed further below, enzymes which remove sufficient extracellular material to cause salivary epithelia to round up into a ball-like mass, do not disperse the microfilaments. Thus, the two effects are distinguishable from each other in a number of respects.

Flat, waterlike salivary epithelia allowed to remain in cytochalasin for as long as 50 hours, can round up and resume normal morphogenesis. By 10 hours after removal of the drug from the culture medium,

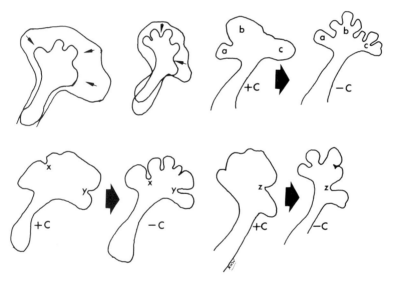

FIG. 16. Tracings of cytochalasin-treated and reversed glands made from projected kodachrome slides of living tissues. In the two examples in the top row, left, the tracings are superimposed upon each other to emphasize the remarkable reduction in area covered by the epithelium after reversal of the cytochalasin effect. In those examples, as well as in the other three pairs, note how new clefts reappear in the 18 hours after drug removal. Areas such as b, upper right, form several lobules; deep clefts, such as x, are not completely eliminated by drug action, and they deepen further during the reversal phase. The differences illustrated here reflect in large part the stage differences of the epithelia at the time of explantation and initial exposure to cytochalasin.

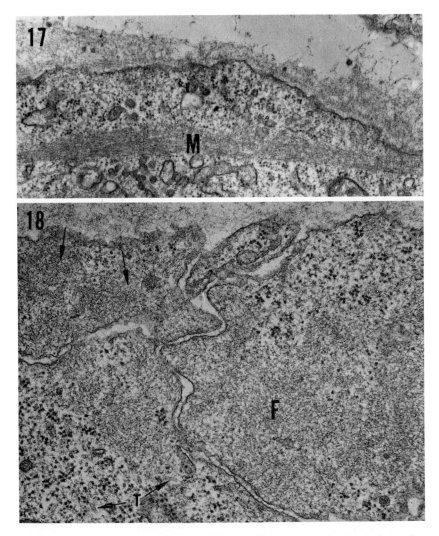

FIG. 17. A bundle of microfilaments (M) extending across the basal cytoplasm of a salivary epithelial cell, as seen in glands fixed immediately after the embryo was removed from the uterus, or at any time of culture under control conditions. × 34,000.

FIG. 18. The basal ends of salivary epithelial cells from a gland cultured in cytochalasin for 18 hours. Bundles of microfilaments as in Fig. 17 are not seen. Instead, only short segments of filamentous material (F) are evident. The lengths of filaments at the upper left (arrows) are representative of the longest filaments found after drug treatment. Note the short lengths of microtubules (T); in other planes of section, very long cytoplasmic microtubules and normal spindle microtubules appear common. × 30,000.

such epithelia begin to look "thicker" as their edges retract. As seen in the superimposed tracings in Fig. 16, a substantial reduction in the area covered by the epithelium takes place 18–24 hours after drug removal. As the rounded configuration is reestablished, clefts reappear at the surface of the epithelium (Figs. 15b and 16). This process also is observed first at ca. 10 hours; by 18–24 hours, deep wide clefts are present. If such recovered glands remain in culture, progressive branching resumes and characteristic salivary epithelial morphology reappears, clearly indicating that cytochalasin does not cause irreparable damage.

Electron micrographs establish that bundles of microfilaments reappear in epithelia that recover from cytochalasin (Fig. 19). The filaments insert in the electron dense material at the cell periphery as they did prior to cytochalasin application. The surface of the salivary gland duct and that of the tips and sides of the lobules are quite smooth; i.e., the plasma membrane appears straight, and the basal lamina is closely applied to its surface in all regions.

Examination of the clefts which appear after removal of cytochalasin reveals extraordinarily large bundles of microfilaments at the basal ends of the cells. Frequently, the plasma membrane at the basal surface of the epithelial cells is thrown into twists and convolutions. The basal lamina is frequently separated from the plasma membrane surface and extends in tortuous whorls into the mesenchymal spaces. In both these respects, the surface at the base of the reformed clefts is distinct from the smooth surface elsewhere on the epithelium.

Reconstitution of microfilaments after cytochalasin removal was examined in the presence of inhibitors of protein synthesis (Spooner and Wessells, 1971). Puromycin (10^{-5} M) proved to be toxic over the long time periods involved in recovery from cytochalasin. Cycloheximide (7×10^{-5} M) reduced incorporation of labeled amino acids into hot, acid-insoluble material by 95%. In cultures fixed from the same experiment, microfilament bundles were again seen extending across the basal cytoplasm of epithelial cells (Fig. 20). The filament bundles appeared to insert in electron dense materials inside the plasma membrane, as is also seen in control cells. Some masses of cytochalasin-induced filamentous material were still present in the cycloheximide-treated cells, indicating that complete reversal of the cytochalasin phenomenon had not occurred (longer time periods for recovery have yet to be afforded such cultures). These results parallel those of Yamada et al. (1970), who found that nerves recover from

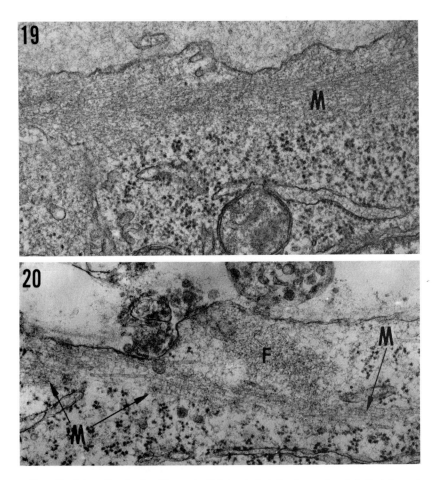

FIG. 19. A bundle of microfilaments (*M*) located at the base of a cleft that formed following removal of cytochalasin. This band of filaments could be traced across five cells at the base of such a cleft. × 45,000.

FIG. 20. Microfilaments (*M*) in the basal cytoplasm of a salivary epithelial cell that recovered from cytochalasin treatment *in the presence of cycloheximide*. Compare this condition with that seen in Fig. 18. Note here, the region of fine filamentous material (*F*) remaining in this recovered cell; it is unclear whether such regions represent bundles of microfilaments cut in cross section, or whether reversal of the drug effect is incomplete, and such regions are remnants of the extensive masses found in all cytochalasin-treated cells. × 40,000.

cytochalasin inhibition when in the presence of cycloheximide and that substantial growth cone activity and axon elongation can take place in the absence of most protein synthesis. The two pieces of data suggest strongly that reversal of the cytochalasin effect involves reconstitution of functional microfilament assemblies, not synthesis of new ones.

These observations can be interpreted as follows: Cytochalasin causes a disruption in the microfilament system of epithelial cells. As a consequence, forces involved in maintaining epithelial shape are inoperative, resulting in loss of both the characteristic morphogenetic contours and the rounded epithelial shape. After removal of cytochalasin, the microfilament systems reconstitute, and, presumably because of their contractility, the epithelium as a whole rounds up. This apparently occurs because of a widespread phenomenon which could result from action of the microfilaments common to all epithelial cells. Presumably such a contraction occurs slowly enough to allow necessary adjustments in cellular and basal laminar surface areas, so that the epithelium retains a smoothly contoured surface. At points where clefts are to form, however, an accentuated contractile phenomenon presumably takes place, resulting in a substantial reduction in the cross-sectional area of a group of epithelial cells. This process forms depressions in the smooth epithelial surface, much as the center of the lens placode sinks inward as the lens cup forms. Furthermore, the contraction apparently occurs with such speed that adjustments in surface area of the basal ends of the cells, and of the overlying basal lamina, cannot take place rapidly enough to maintain a flat surface. The result is marked folding of the cell surface and actual separation of the basal lamina from the tissue surface. Interestingly, the same foldings and separations have been observed in clefts of normal salivary gland (Fig. 1) and lung as those organs undergo morphogenesis *in vivo*. All these events are most easily explained by assuming that the force responsible is the contractility of the microfilaments.

It is pertinent to ask whether the epithelial changes following cytochalasin application or removal are indirect results of effects upon the mesodermal cell population, since cytochalasin does affect the mesenchyme. In contrast to control cultures, glands grown in cytochalasin do not show migration of individual fibroblasts from the mass of condensed mesenchyme surrounding the epithelium (Fig. 15). Carter (1967) also observed cessation of fibroblast movement after cytochalasin was applied to fibroblast cultures. Such results are predictable, since cytochalasin alters the filamentous network found in the undulating membrane of motile glia cells (Spooner, Yamada, and Wessells, 1971) as well as of the growth cone of elongating axons (Yamada et al., 1970). That other, less apparent, cytochalasin effects upon mesoderm are not the source of epithelial changes is suggested by two kinds of experiments. Transfilter cultures can be prepared in which salivary epithelium continues the salivary-like morphogenesis described by Grobstein and Cohen (1965). Such epithelia become thin and lose their clefts and lobules in response to cytochalasin. When the drug is removed, the epithelia round up as in cultures of intact glands, and at least some branch points reappear. This occurs under two circumstances; first, without direct mesenchymal cell proximity to the epithelium (i.e., in normal transfilter culture), and second in cases in which *all* mesenchymal cells were removed from the top of the filter. In the latter instance, recovery occurs even in absence of continuing mesenchymal contribution. These results establish that the recovery process takes place because of forces generated within the epithelial cell population and imply that the analogous changes in epithelial shape during normal morphogenesis are also caused by agents within the epithelium.

The effects of cytochalasin upon salivary glands contrast with those that follow application of agents that disperse microtubules or remove extracellular materials. Colchicine inhibits growth of the epithelium but does not cause either flattening or loss of morphogenetic contours. Thus, 24 hours after treatment with colchicine (2.5×10^{-7} M) is started, the epithelium is approximately the same size as at zero time and clefts and lobules present at explantation are still intact. Electron micrographs show that both spindle and cytoplasmic microtubules are absent from the cytoplasm of treated cells, but that both 50 Å microfilaments and 100 Å desmosome-associated filaments remain unaltered in morphology. Interestingly, epithelial cells located on the periphery of the tissue appear cuboidal or rather irregular in shape—thus, reflecting dissolution of microtubules—but this has no noticeable effect upon tissue shape: lobules and clefts remain. This observation parallels that of Pearce and Zwaan (1970), who failed to see striking alterations in lens tissue shape or morphogenesis despite Colcemid treatment.

Since experiments cited above demonstrate that various enzyme treatments can produce striking alterations in epithelial morphology and morphogenesis, it is of interest to determine the effect of such enzymes upon intracellular organelles. In tissues fixed directly after treatment with collagenase, collagenase followed by hyaluronidase, or trypsin-pancreatin, microfilaments and microtubules remain intact within epithelial cells (Figs. 10 and 11). Microfilaments occur in bundles and insert in an apparently normal manner at the cell periphery. Furthermore, in epithelia treated with collagenase and hyaluronidase and cultured for 17 hours in the presence of salivary mesoderm, microfilament bundles also appear normal at the basal ends of the epithelial cells (Figs. 12 and 13). As pointed out above, such epithelia are in a temporary state of "disorganization" and only later does recognizable morphogenesis start anew. Also of note is that such disorganized epithelia are better described as being "rounded into a ball" than as being in the thinned condition following culture in cytochalasin. The presence of microfilaments under such conditions is predictable if in addition to being involved in morphogenetic events at certain points in the epithelium, the hypothesis that microfilaments are involved in maintenance of a rounded epithelial shape is correct.

Before considering the role of microfilaments in other systems, it should be emphasized again that microfilament function in morphogenesis is probably a transient phenomenon. Deep clefts in a salivary surface are in part stable in cytochalasin (Figs. 16 and 22); as will be described below, early phases of tubular gland formation in oviduct are reversed if microfilaments are disrupted, whereas older glands are stable. In such situations, accumulation of extracellular materials, of large number of cells, or other processes may be sources of stabilization that cannot be overcome by microfilament disruption.

The action of cytochalasin as an inhibitor of morphogenesis is not restricted to the salivary gland. Chick oviduct responds to estradiol injection by forming microfilament assemblies and early tubular glands (see above; Wrenn and Wessells, 1970). Such an oviduct, placed in cytochalasin for 6 hours, shows complete inhibition of new gland formation, and disappearance of the small glands already present. The latter phenomenon is of particular interest, for there is every reason to presume that the contractile activity of the microfilament bundles generated the early glandular evaginations. Just as is the case in salivary epithelial cells, cytochalasin disrupts the microfilament assemblies of oviduct cells, leaving in their place masses or aggregates of short filaments. Microtubules, and other organelles appear normal after drug treatment. The implication from these observations is of course that the actual integrity of the early evagination, its lumen in particular, is dependent upon presence of the microfilament contractility. As is the case with salivary tissues, microfilaments appear once more after cytochalasin is removed from the media bathing the oviduct tissue.

Effects of cytochalasin on developing lung are identical to those on salivary and oviducal epithelia in that morphogenesis of bronchi is arrested. However, despite the fact that microfilaments are disrupted, interpretation of the result is complicated by the extraordinarily high rate of cell division in the tracheobronchial tree (Wessells, 1970). As predicted from Carter's (1967) experiments (see also Ridler and Smith, 1968), multinucleate cells accumulate in the lumen of the lung system when cytochalasin is present. These multinucleate cells apparently do not migrate to the basal surface of the epithelium as is the case following normal cell division. Often the multinucleate cells show signs of ill-health and necrosis. As a result of these processes, continued culture with cytochalasin yields a single, thin layer of healthy cells which outline the epithelium of the lung. Because of this effect it is impossible to say that the cessation of bronchial morphogenesis is due solely to disruption of microfilaments. Nevertheless, the lung system is consistent with salivary, oviduct, and all others examined in providing positive correlations between presence of intact microfilaments with morphogenesis, and disruption of filaments with cessation of morphogenesis.

Multinucleate cells also appear in salivary and oviducal epithelia cultured in cytochalasin. Their number is not sufficiently great to generate derangements in epithelial structure. Furthermore, such cells fail to interfere with recovery from cytochalasin, and in the case of salivary, with continuing morphogenesis thereafter.

Probably the strongest argument that the primary effects of cytochalasin are not caused by inhibition of cytokinesis and accumulation of multinucleate cells comes from work on *nondividing* nerve cells (Yamada et al., 1970). Nerve cells dissociated from chick embryo spinal ganglia or ventral halves of spinal cord can be grown in cell culture in the presence of scattered glia cells. Axons elongate from both types of nerve cells, in the former case at a speed of 20–40 μ per hour. When cytochalasin is added to the culture medium, growth cone and microspike activity at the tip of each advancing axon ceases within several minutes. Axon elongation halts concomitantly. Glial cells also cease movement very rapidly and shrink into a stellate shape. In the latter case, masses of fine filamentous material, indistinguishable from that in salivary and oviducal epithelial cells, appear in the cytoplasm of such cells. There is no apparent effect of cytochalasin on neurotubules and 100 Å neurofilaments in nerve cells. The only observed alteration is in a network of 50 Å filamentous material that is the sole organelle found in the microspikes and the peripheral growth cone. The drug causes a "condensation" of the network so that individual filament segments appear more tightly packed than in control axon tips. When cytochalasin is washed from the culture dishes, within one to several hours, microspikes reappear, growth cone activity starts anew, and axon elongation commences again.

One feature of these cases is consistent: the integrity of microfilament systems is required for multicellular or single cell morphogenesis. Furthermore, properties such as cytokinesis and cell movement that might be thought of as being of a general cellular nature, as opposed to being developmental in nature, are also dependent upon the integrity of microfilaments.

These data provide evidence that a class of contractile organelles is present in a wide variety of cells and that those organelles can be mobilized for a number of distinct functions. Periodic assembly and function of the contractile ring microfilaments, or of the undulating membrane filaments of migratory cells, might be examples of general use of microfilaments. Assembly and function at one end of an epithelial cell, at a particular time, is an example of a developmental role for the structures. While little can said of the microfilaments, at least their function and morphological integrity seem affected by the drug cytochalasin. The parallel to colchicine effects upon different sorts of microtubules is obvious. In the cytochalasin experiments, however, one does not yet know whether the microfilaments themselves are dispersed by the drug, or whether the insertion points of the microfilament assembly upon the cell surface is the site of action.

The considerations of Tilney (1968) with respect to microtubules in developing cells apply equally to microfilaments. At least two variables must be under developmental control for organelles within individual cells to be used in a manner that causes cell populations to change shape in a coordinated way. The first is the means by which the microfilament systems are assembled in particular places at particular times in cells. Baker and Schroeder (1967) pointed out, for instance, that only cells near the midline of the medullary plate —destined to be the floor of the neural tube—acquire filament bundles at their outer ends. Thus, the assembly process is not only controlled in time and space within a cell but also within a cell population. The second variable is the control of microfilament function. If, in a given population of cells all possessing microfilaments, contraction occurs at one time in cell x, 6 hours later in cell y, and so on, the population as a whole would not necessarily show the results of group contraction. In other words the medullary plate would not depress along its midline if scattered cells showed the pursestring contraction, as opposed to the case in which all cells in the population contracted.

The means by which organelle distribution and function is controlled in epithelial cells are of paramount concern in developing organs such as salivary gland. If the mesoderm, or its extracellular secretions (see above), plays a pivotal role in epithelial morphogenesis, then it becomes crucial to know whether that effect is a direct one upon the organelle machinery, or whether it is indirect via the nucleus or genes of the epithelial cell. Operationally, the two possibilities cannot be distinguished at the moment. It is possible for example that the salivary mesoderm, in its very presence, provides conditions in which a set of genes within salivary epithelial cells carries out the whole developmental program leading to salivary development: that is, genes involved in organelle synthesis, assembly, and function, those involved in interaction with the mesoderm, and those concerned with cytodifferentiation all are brought into action as a group. Under such circumstances the mesoderm might in a sense provide the physical framework within which the epithelium branches, as well as cues enabling the epithelial genes to operate.

As an alternative, the mesoderm might play a role in epithelial morphogenesis by affecting the microfilament and microtubule systems directly. As pointed out above, all salivary epithelial cells possess microfilaments that are apparently responsible for the rounded tissue shape. At predictable, presumably genetically determined, points within the epithelial population, an accentuated contraction phenomenon apparently occurs to produce or to stabilize the "clefts" of the epithelium. Mesoderm might be involved in causing extraordinary numbers or kinds of microfilaments to form in those epithelial cells destined to lie at the base of clefts. Or, mesoderm might control the activity of the microfilaments, thus leading to cleft formation.

The latter hypothesis provides a means of explaining how mouse lung epithelium can "branch" initially in an avian pattern when combined with avian lung mesoderm (Taderera, 1967). It is difficult to imagine in that case why murine cells would possess genetic information that could lead to an avian bronchial branching pattern. Direct mesodermal influence on a malleable epithelial population seems more attractive there, as well as in the cases where salivary-like morphogenesis is elicited from mammary (Kratochwil, 1969), or bronchial epithelia (Spooner and Wessells, 1970b).

Despite the attractiveness of the hypothesis that mesoderm effects epithelial morphogenetic machinery per se, there are major unexplained difficulties with the concept. The main problem that must be faced was discussed by Harrison (1933) with respect to determination: that is, given the proposition of localized mesodermal control activity, what distinguishes one mesodermal subpopulation from another? That presumably is essential if one subpopulation is to lead to cleft formation by the epithelium, another to be associated with a rounded lobule, or another with a duct.

CONCLUDING REMARKS

We have described experiments in which extra- and intracellular factors were studied for their effect on the maintenance and formation of epithelial morphology. These studies, which indicate that epithelial morphogenesis is dependent upon both surface-associated mucopolysaccharide–protein complexes and contractile filamentous organelles, provide the following evidence: Both factors are identified in normal, untreated glands; disruption or loss of either factor correlates with loss of characteristic morphology, while reappearance of that factor is associated with the resumption of morphogenesis. The data suggest that the functions of the factors are complementary in that removal of the surface-associated material results in a round, "ball-like" epithelium, while disruption of the microfilaments yields a flat, "waferlike" rudiment (Fig. 21). Similarly, recovery of morphology in the former case occurs by the formation of outgrowths from the spherical epithelium, which subsequently undergo progressive branching, while reversal of the cytochalasin effect leads to the reappearance of deep clefts concomitant with "thickening" of the epithelium. Removal of surface-associated mucopolysaccharide–

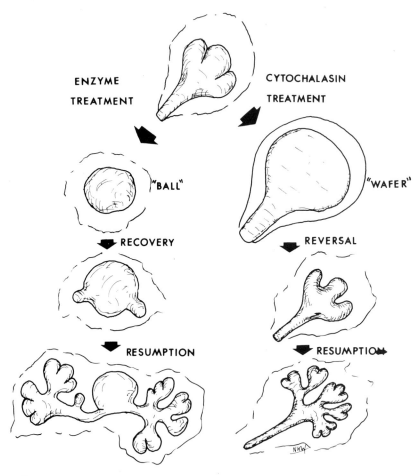

Fig. 21. A schematic representation emphasizing the differences in salivary epithelial behavior following removal of mucopolysaccharide-protein from the surface *versus* disruption of microfilament function. In the former situation, the epithelium rounds up and sprouts new outgrowths, each of which undergoes morphogenesis much like that of a whole salivary gland. In the latter case, the epithelium flattens and loses clefts; during reversal, clefts form and the epithelium thickens.

protein did not result in loss of microfilaments, and disruption of the microfilaments did not cause a loss of extracellular material, suggesting that these factors operate independently, but are not sufficient by themselves to maintain epithelial shape and to promote morphogenesis.

Since morphogenesis proceeds in a precise and highly ordered fashion, the factors involved in this process must be closely integrated. From the studies presented in this paper, a hypothetical model for the interrelationship of morphogenetically active surface-associated material and of the putatively contractile microfilaments can be formulated. This model accounts for all the available data and makes predictions which are subject to experimental verification. The model is based upon the following premises: (1) the microfilaments in the cells at the basal surface of the epithelium are contractile; (2) mucopolysaccharide–protein complexes are present near the epithelial surface; (3) collagen fibers are also present near the epithelial surface, and arise by the interaction of collagen (synthesized predominantly by mesenchyme) with the mucopolysaccharide–protein complexes; and (4) regions of accentuated microfilament contractility, which result in cleft formation in an epithelium, are caused by localized areas of surface-associated mucopolysaccharide–protein.

The model may be used to explain the results of cytochalasin treatment and of removal of surface-associated mucopolysaccharide–protein from salivary epithelia. The loss of morphology, and formation of a "waferlike" epithelium after the addition of cytochalasin to epithelial cultures could be due to cessation of the contractile force of the microfilaments. Removal of surface-associated mucopolysaccharide–protein results in loss of morphology, and formation of a "ball-like" epithelium containing basal epithelial cells which have lost their alignment with respect to the tissue surface. The epithelium assumes a spherical shape possibly because the lack of mucopolysaccharide–protein in the extracellular matrix removes the stimulus for accentuated local microfilament activity.

During reversal of cytochalasin treatment, microfilament activity presumably resumes, and, because the surface-associated materials are retained, localized contractions leading to cleft formation take place rapidly. During recovery from the loss of surface mucopolysaccharide–protein, newly synthesized materials are presumably laid down at various sites, thus producing local areas where microfilament contractility and possibly mitosis, are accentuated. These processes yield an evagination of cells which resume branching morphogenesis.

What is the relationship between this model, based upon the foregoing experimental studies, and normal morphogenesis? The distribution on the epithelial surface of newly synthesized acid mucopolysaccharide, which contrasts with that of total acid mucopolysaccharide and that of collagen fibers seen ultrastructurally, may provide an insight into this relationship. Since, according to the model, surface-associated mucopolysaccharide is thought to be the stimulus for accentuated microfilament activity, the formation of clefts should be associated with localized accumulations of newly synthesized mucopolysaccharide. Areas at the epithelial surface near the distal ends of the growing lobules, the sites of incipient cleft formation, are indeed the areas where the greatest amount of radioactive mucopolysaccharide is seen after a brief labeling period. On the other hand, clefts that presumably resulted from prior microfilament activity, are characterized by considerably less incorporation of mucopolysaccharide precursors. The clefts and the surface of the lobules show similar amounts of mucopolysaccharide as revealed by specific stain-

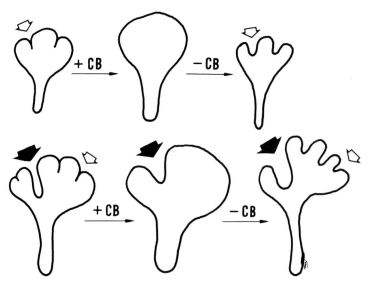

Fig. 22. A comparison of cytochalasin treatment of "young" versus "old" 13-day salivary glands. The younger gland (top) possesses only narrow, recently formed clefts (arrow). All such clefts disappear after incubation in cytochalasin. Clefts then reappear after drug removal. The older gland (below) contains one deep, wide cleft (solid arrow) and several narrow ones (arrow) analogous to those seen in the younger gland. The narrow clefts disappear and reappear with cytochalasin application and removal; the deep cleft does not disappear. The latter behavior would be expected if clefts are stabilized by collagen fibers that accumulate in a time-dependent manner. The interpretation is consistent with the idea that microfilaments are involved in formation and stability of early clefts, but that extracellular materials play an essential role in stabilization at later times.

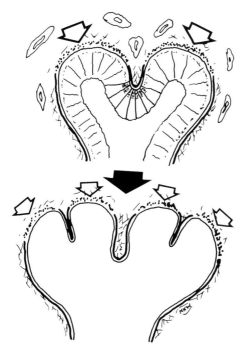

FIG. 23. A model showing the relationship of incipient cleft formation to sites of accumulation of extracellular materials. The outer surfaces of the epithelium and the internal surfaces of the deep cleft are sites of accumulation of collagen and of low rates of acid mucopolysaccharide deposition. The tips of the lobules, near the points of incipient cleft formation (open arrows), have less surface-associated substances and more rapid deposition of newly synthesized mucopolysaccharide. As clefts form (lower figure), extracellular materials accumulate within the clefts due to interaction of acid mucopolysaccharide-protein with collagen. Concomitantly there is a decrease in the rate at which newly synthesized mucopolysaccharide is deposited. Areas adjacent to the new clefts retain the original properties, and, after suitable growth of the lobules, serve as sites of subsequent cleft formation.

ing, but the material in the clefts probably was synthesized at the time when cleft formation took place. This concept is supported by the distribution of collagen fibers. Collagen fibrogenesis is thought to occur as a result of tropocollagen interaction with acid mucopolysaccharide–protein complexes (Toole, 1969). The greater distribution of collagen fibers at morphogenetically quiescent areas (i.e., clefts, interbronchial areas) may reflect the earlier presence of mucopolysaccharide–protein at these sites. Thus, these observations, made on intact glands, can be understood in terms of the mechanism proposed for resumption of morphogenesis after cytochalasin treatment or after removal of surface-associated mucopolysaccharide–protein. Subsequent studies will, hopefully, shed light on the validity of this hypothesis.

ACKNOWLEDGMENT

We are pleased to thank our colleagues Shib D. Banerjee, Brian S. Spooner, Joan T. Wrenn, and Kenneth M. Yamada for their stimulating participation in these studies. Work reported here was supported by Public Health Service Research Grants HD-02147, GM-16530, HD-04708, National Science Foundation Grant GB-15666, and National Foundation-March of Dimes Grant CRCS-40.

REFERENCES

ALESCIO, T., and CASSINI, A. (1962). Induction *in vitro* of tracheal buds by pulmonary mesenchyme grafted on tracheal epithelium. *J. Exp. Zool.* **150**, 83–94.

ARNOLD, J. M. (1968). An analysis of cleavage furrow formation in the egg of *Loligo pealii*. *Biol. Bull.* **135**, 413.

AUERBACH, R. (1960). Morphogenetic interactions in the development of the mouse thymus gland. *Develop. Biol.* **2**, 271–284.

BAKER, P. C. (1965). Fine structure and morphogenetic movements in the gastrula of the tree frog, *Hyla regilla*. *J. Cell Biol.* **24**, 95–116.

BAKER, P. C., and SCHROEDER, T. E. (1967). Cytoplasmic filaments and morphogenetic movement in the amphibian neural tube. *Develop. Biol.* **15**, 432–450.

BERLINER, J. (1969). The effects of the epidermis on the collagenous basement lamella of anuran larval skin. *Develop. Biol.* **20**, 544–562.

BERNFIELD, M. R. (1970). Collagen synthesis during epitheliomesenchymal interactions. *Develop. Biol.* **22**, 213–231.

BERNFIELD, M. R., and BANERJEE, S. D. (1971a). Acid mucopolysaccharide at the epithelial-mesenchymal interface of mouse embryo salivary glands. In preparation.

BERNFIELD, M. R., and BANERJEE, S. D. (1971b). Dependence of epithelial morphology and morphogenesis upon extracellular acid mucopolysaccharide–protein at the epithelial surface. In preparation.

BORGHESE, E. (1950). The development *in vitro* of the submandibular and sublingual glands of *Mus musculus*. *J. Anat.* **84**, 287–302.

BYERS, B., and PORTER, K. R. (1964). Oriented microtubules in elongating cells of the developing lens rudiment after induction. *Proc. Nat. Acad. Sci. U.S.* **52**, 1091–1099.

CARTER, S. B. (1967). Effects of cytochalasin on mammalian cells. *Nature (London)* **213**, 261–264.

CLONEY, R. A. (1966). Cytoplasmic filaments and cell movements: epidermal cells during ascidian metamorphosis. *J. Ultrastruct. Res.* **14**, 300–328.

CLONEY, R. A. (1969). Cytoplasmic filaments and morphogenesis: the role of the notochord in ascidian metamorphosis. *Z. Zellforsch. Mikrosk. Anat.* **100**, 31–53.

COOK, G. M. W. (1968). Glycoproteins in membranes. *Biol. Rev. Cambridge Phil. Soc.* **43**, 363–391.

DAMERON, F. (1968). Etude expérimentale de l'organogenèse du poumon: nature et spécificité des interaction épithelio-mésenchymateuses. *J. Embryol. Exp. Morphol.* **20**, 151–167.

Davis, N. C., and Smith, E. L. (1955). Assay of proteolytic enzymes. In "Methods of Biochemical Analysis" (D. Glick, ed.), Vol. 2. Wiley (Interscience), New York.
Dodson, J. W. (1967). Differentiation of epidermis. I. Interrelationship of epidermis and dermis in embryonic chicken skin. *J. Embryol. Exp. Morphol.* **17**, 83–106.
Edds, M. V., and Sweeny, P. R. (1961). Chemical and morphological differentiation of the basement lamella. In "Synthesis of Molecular and Cellular Structure" (D. Rudnick, ed.) pp. 111–138. Ronald Press, New York.
Farquhar, M. G., and Palade, G. E. (1963). Functional complexes in various epithelia. *J. Cell Biol.* **17**, 375–412.
Fitton Jackson, S. (1968). The morphogenesis of collagen. In *"Treatise on Collagen"* Vol. 2, "Biology of Collagen" (B. S. Gould, ed.), Part B, pp. 1–66. Academic Press, New York.
Goldberg, B., and Green, H. (1968). The synthesis of collagen and protocollagen hydroxylase by fibroblastic and non-fibroblastic cell lines. *Proc. Nat. Acad. Sci. U.S.* **59**, 1110–1115.
Golosow, N., and Grobstein, C. (1962). Epitheliomesenchymal interaction in pancreatic morphogenesis. *Develop. Biol.* **4**, 242–255.
Grobstein, C. (1953). Epithelio-mesenchymal specificity in the morphogenesis of mouse sub-mandibular rudiments *in vitro. J. Exp. Zool.* **124**, 383–404.
Grobstein, C. (1967). Mechanism of organogenetic tissue interaction. *Nat. Cancer Inst. Monogr.* **26**, 279–299.
Grobstein, C., and Cohen, J. (1965). Collagenase: Effect on the morphogenesis of embryonic salivary epithelium *in vitro. Science* **150**, 626–628.
Harper, E., Seifter, S., and Hospelkorn, V. D. (1965). Evidence for subunits in bacterial collagenase. *Biochem. Biophys. Res. Commun.* **18**, 627–633.
Harrison, R. G. (1933). Some difficulties of the determination problem. *Amer. Naturalist* **67**, 306–321.
Hay, E. D., and Revel, J. P. (1963). Autoradiographic studies of the origin of the basement lamella in *Ambystoma. Develop. Biol.* **7**, 303–323.
Hay, E. D., and Revel, J. P. (1969). Fine structure of the developing avian cornea. In "Monographs in Developmental Biology" (A. Wolsky and P. S. Chen, eds.), pp. 16–46. Karger, New York.
Holtfreter, J. (1968). Mesenchyme and epithelia in inductive and morphogenetic processes. In *"Epithelial-Mesenchymal Interactions"* (R. Fleischmajer and R. E. Billingham, eds.) (*Hahnemann Symp. 18*), pp. 1–30. Williams and Wilkins, Baltimore, Maryland.
Hummel, B. C. W. (1959). A modified spectrophotometric determination of chymotrypsin, trypsin, and thrombin. *Can. J. Biochem. Physiol.* **37**, 1393–1399.
Ito, S. (1969). Structure and function of the glycocalyx. *Fed. Proc., Fed. Amer. Soc. Exp. Biol.* **28**, 12–25.
Janners, M. V., and Searls, R. L. (1968). Changing growth rates in the mesoderm of the embryonic chick wing bud. *J. Cell Biol.* **39**, 67a.
Kallman, F., and Grobstein, C. (1964). Fine structure of differentiating mouse pancreatic exocrine cells in transfilter culture. *J. Cell Biol.* **20**, 399–413.
Kallman, F., and Grobstein, C. (1965). Source of collagen at epitheliomesenchymal interfaces during inductive interaction. *Develop. Biol.* **11**, 169–183.
Kallman, F., and Grobstein, C. (1966). Localization of glucosamine-incorporating materials at epithelial surfaces during salivary epitheliomesenchymal interaction *in vitro. Develop. Biol.* **14**, 52–67.

KALLMAN, F., EVANS, J., and WESSELLS, N. K. (1967). Normal epidermal basal cell behavior in the absence of basement membrane. *J. Cell Biol.* **32,** 231–236.

KEFALIDES, N. A. (1968). Isolation and characterization of the collagen from glomerular basement membrane. *Biochemistry* **7,** 3103–3112.

KEFALIDES, N. A., and WINZLER, R. J. (1966). The chemistry of glomerular basement membrane and its relation to collagen. *Biochemistry* **5,** 702–713.

KELLY, D. E. (1966). Fine structure of desmosomes, hemidesmosomes, and an adepidermal globular layer in developing newt epidermis. *J. Cell Biol.* **28,** 51–72.

KRAEMER, P. M. (1966). Sialic acid of mammalian cell lines. *J. Cell Physiol.* **67,** 23–34.

KRATOCHWIL, K. (1969). Organ specificity in mesenchymal induction demonstrated in the embryonic development of the mammary gland of the mouse. *Develop. Biol.* **20,** 46–71.

LENT, R., and SEIFTER, S. (1969). Personal communication.

LILLIE, R. D. (1965). "Histopathologic Technic and Practical Histochemistry," 3rd ed., pp. 198–202. McGraw-Hill, New York.

LINKER, A. (1966). Bacterial mucopolysaccharidases (mucopolysaccharide lyases). *In* "Methods in Enzymology" (S. P. Colowick and N. O. Kaplan, eds.), Vol. VIII, pp. 650–654. Academic Press, New York.

MCINTOSH, J. F., and PORTER, K. R. (1967). Microtubules in the spermatids of the domestic fowl. *J. Cell Biol.* **35,** 153–174.

MATHEWS, M. B. (1965). The interaction of collagen and acid mucopolysaccharides. *Biochem. J.* **96,** 710–716.

MATHEWS, M. B. (1966). Animal mucopolysaccharidases. *In* "Methods in Enzymology" (S. P. Colowick and N. O. Kaplan, eds.), Vol. VIII, pp. 654–662. Academic Press, New York.

MAYHEW, E., and WEISS, L. (1968). Ribonucleic acid at the periphery of different cell types and effect of growth rate on ionogenic groups in the periphery of cultured cells. *Exp. Cell Res.* **50,** 441–453.

NADOL, J. P., JR., GIBBINS, J. R., and PORTER, K. R. (1969). A reinterpretation of the structure and development of the basement lamella: an ordered array of collagen in fish skin. *Develop. Biol.* **20,** 304–331.

PEARCE, T. L., and ZWAAN, J. (1970). A light and electron microscopic study of cell behavior and microtubules in the embryonic chicken lens using colcemid. *J. Embryol. Exp. Morphol.* **23,** 491–507.

PIERCE, G. B. (1966). The development of basement membranes of the mouse embryo. *Develop. Biol.* **13,** 231–249.

QUINTARELLI, G. (1968). Methods for the histochemical identification of acid mucopolysaccharides. A critical evaluation. *In* "The Chemical Physiology of Mucopolysaccharides" (G. Quintarelli, ed.), pp. 199–218. Little, Brown, Boston, Massachusetts.

QUINTARELLI, G., and DELLOVO, M. C. (1965). The chemical and histochemical properties of Alcian blue. IV. Further studies on the methods for the identification of acid glycosaminoglycans. *Histochemie* **5,** 196–209.

QUINTARELLI, G., SCOTT, J. E., and DELLOVO, M. C. (1964). The chemical and histochemical properties of Alcian blue. II. Dye binding of tissue polyanions. *Histochemie* **4,** 86–98.

RAMBOURG, A., and LEBLOND, C. P. (1967). Electron microscope observations on the carbohydrate-rich cell coat present at the surface of cells in the rat. *J. Cell Biol.* **32,** 27–53.

Rambourg, A., Neutra, M., and Leblond, C. P. (1966). Presence of a "cell coat" rich in carbohydrate at the surface of cells in the rat. *Anat. Rec.* **154,** 41–71.

Rappaport, R. (1967). Cell division: direct measurement of maximum tension exerted by furrow of Echinoderm eggs. *Science* **156,** 1241–1243.

Ridler, M. A. C., and Smith, G. F. (1968). The response of human cultured lymphocytes to cytochalasin B. *J. Cell Sci.* **3,** 595–602.

Rifkind, R. A., Chui, D., and Epler, H. (1969). An ultrastructural study of early morphogenetic events during the establishment of fetal hepatic erythropoiesis. *J. Cell Biol.* **40,** 343–365.

Rodén, L. (1968). Linkage of acid mucopolysaccharides to protein. In "Fourth International Conference on Cystic Fibrosis of the Pancreas," Part II: "Biochemistry of Glycoproteins and Related Substances" (E. Rossi and E. Stoll, eds.), pp. 185–202. Karger, New York.

Ronzio, R. A., and Rutter, W. J. (1969). The role of mesodermal factor in pancreatic differentiation: Stimulation of DNA synthesis. *J. Cell Biol.* **43,** 118a–119a.

Rutter, W. J., Wessells, N. K., and Grobstein, C. (1964). Control of specific synthesis in the developing pancreas. *Nat. Cancer Inst. Monogr.* **13,** 51–65.

Saunders, J. W., Jr. (1966). Death in embryonic systems. *Science* **154,** 604–612.

Schroeder, T. E. (1968). Cytokinesis: filaments in the cleavage furrow. *Exp. Cell Res.* **53,** 272–276.

Schroeder, T. E. (1969). The role of "contractile ring" filaments in dividing *Arbacia* eggs. *Biol. Bull.* **137,** 413.

Scott, J. E., and Dorling, J. (1965). Differential staining of acid glycosaminoglycans (mucopolysaccharides) by Alcian blue in salt solutions. *Histochemie* **5,** 221–233.

Scott, J. E., and Dorling, J. (1969). Periodate oxidation of acid polysaccharides. III. A PAS method for chondroitin sulphates and other glycosamino-glycuronans. *Histochemie* **19,** 295–301.

Scott, J. E., and Harbinson, R. J. (1969). Periodic oxidation of acid polysaccharides. II. Rates of oxidation of uronic acids in polyuronides and acid mucopolysaccharides. *Histochemie* **19,** 155–161.

Scott, J. E., Quintarelli, G., and Dellovo, M. C. (1964). The chemical and histochemical properties of Alcian blue. I. The mechanism of Alcian blue staining. *Histochemie* **4,** 73–85.

Seifter, S., Gallop, P. M., Klein, L., and Meilman, E. (1959). Studies on collagen. II. Properties of purified collagenase and its inhibition. *J. Biol. Chem.* **234,** 285–293.

Slavkin, C., Bringas, P., Cameron, J., LeBaron, R., and Bavetta, L. A. (1969b). actions *in vitro*. *J. Dent. Res.* **47,** 779–785.

Slavkin, H. C., Bringas, P., and Bavetta, L. A. (1969a). Ribonucleic acid within the extracellular matrix during embryonic tooth formation. *J. Cell Physiol.* **73,** 179–190.

Slavkin, H. C., Bringas, P., Cameron, J., LeBaron, R., and Bavetta, L. A. (1969b). Epithelial and mesenchymal cell interactions with extracellular matrix material *in vitro*. *J. Embryol. Exp. Morphol.* **22,** 395–405.

Spiro, R. G. (1967). Studies on the renal glomerular basement membrane. Nature of the carbohydrate units and their attachment to the peptide portion. *J. Biol. Chem.* **242,** 1923–1932.

Spiro, R. G., and Fukushi, S. (1969). The lens capsule: Studies on the carbohydrate units. *J. Biol. Chem.* **244,** 2049–2058.

Spooner, B. S., and Wessells, N. K. (1970a). Effects of cytochalasin B upon microfilaments involved in morphogenesis of salivary epithelium. *Proc. Nat. Acad. Sci. U.S.* **66**, 360–364.

Spooner, B. S., and Wessells, N. K. (1970b). Mammalian lung development: interactions in primordium formation and bronchial morphogenesis. *J. Exp. Zool.* **175**, 445–454.

Spooner, B. S., and Wessells, N. K. (1971). An analysis of salivary gland morphogenesis. In preparation.

Spooner, B. S., Yamada, K. M., and Wessells, N. K. (1971) Microfilaments and Cell Locomotion. *J. Cell Biol.* in press.

Stuart, E. S., and Moscona, A. A. (1967). Embryonic morphogenesis: Role of fibrous lattice in the development of feathers and feather patterns. *Science* **157**, 947–948.

Szollosi, D. (1970). Cortical cytoplasmic filaments of cleaving eggs: a structural element corresponding to the contractile ring. *J. Cell Biol.* **44**, 192–209.

Taderera,, J. V. (1967). Control of lung differentiation *in vitro*. *Develop. Biol.* **16**, 489–512.

Thorp, F. K., and Dorfman, A. (1967). Differentiation of connective tissues. *Curr. Top. Develop. Biol.* **2**, 151–190.

Tilney, L. G. (1968). II. Ordering of subcellular units. The assembly of microtubules and their role in the development of cell form. *Develop. Biol.*, Suppl. 2, 63–102.

Toole, B. P. (1969). Solubility of collagen fibrils formed *in vitro* in the presence of sulphated acid mucopolysaccharide–protein. *Nature (London)* **222**, 872–873.

Toole, B. P., and Lowther, D. A. (1968a). The effect of chondroitin sulphate-protein on the formation of collagen fibrils *in vitro*. *Biochem. J.* **109**, 857–866.

Toole, B. P., and Lowther, D. A. (1968b). Dermatan sulfate-protein: Isolation from and interaction with collagen. *Arch. Biochem. Biophys.* **128**, 567–578.

Weiss, L., and Mayhew, E. (1966). The presence of ribonucleic acid within the peripheral zones of two types of mammalian cell. *J. Cell Physiol.* **68**, 345–360.

Wessells, N. K. (1964). Substrate and nutrient effects upon epidermal basal cell orientation and proliferation. *Proc. Nat. Acad. Sci. U.S.* **52**, 252–259.

Wessells, N. K. (1968). Problems in the analysis of determination, mitosis, and differentiation. "Epithelial-Mesenchymal Interactions" (R. Fleischmajer and R. E. Billingham, eds.) (*Hahnemann Symp. 18*), pp. 132–151. Williams and Wilkins, Baltimore, Maryland.

Wessells, N. K. (1970). Mammalian lung development: interactions in formation and morphogenesis of tracheal buds. *J. Exp. Zool.*, **175**, 455–466.

Wessells, N. K., and Cohen, J. H. (1966). The influence of collagen and embryo extract on the development of pancreatic epithelium. *Exp. Cell Res.* **43**, 680–684.

Wessells, N. K., and Cohen, J. H. (1968). Effects of collagenase on developing epithelia *in vitro*: lung, ureteric bud, and pancreas. *Develop. Biol.* **18**, 294–309.

Wessells, N. K., and Evans, J. (1968a). Ultrastructural studies of early morphogenesis and cytodifferentiation in the embryonic mammalian pancreas. *Develop. Biol.* **17**, 413–446.

Wessells, N. K., and Evans, J. (1968b). The ultrastructure of oriented cells and extracellular materials between developing feathers. *Develop. Biol.* **18**, 42–61.

Weston, J. A. (1970). The migration and differentiation of neural crest cells. *Advan. Morphog.* **8**, 41–114.

Wolff, E. (1968). Specific interactions between tissues during organogenesis. *Curr. Top. Develop. Biol.* **3**, 65–94.

WRENN, J. T., and WESSELLS, N. K. (1969). An ultrastructural study of lens invagination in the mouse. *J. Exp. Zool.* **171,** 359–368.
WRENN, J. T., and WESSELLS, N. K. (1970). Cytochalasin B: effects upon microfilaments involved in morphogenesis of estrogen-induced glands of oviduct. *Proc. Nat. Acad. Sci. U.S.* **66,** 904–908.
WU, H. C., MEEZAN, E., BLACK, P. H., and ROBBINS, P. W. (1969). Comparative studies on the carbohydrate-containing membrane components of normal and virus transformed mouse fibroblasts. I. Glucosamine labeling patterns in 3T3, spontaneously transformed 3T3, and SV-40-transformed 3T3 cells. *Biochemistry* **8,** 2509–2517.
WÜNSCH, E., and HEIDRICH, H. (1963). Zurqualitativen bestimmung der Kollagenase. *Hoppe Seyler's Z. Physiol. Chem.* **333,** 149–151.
YAMADA, K. M., SPOONER, B. S., and WESSELLS, N. K. (1970). Axon growth: roles of microfilaments and microtubules. *Proc. Nat. Acad. Sci. U.S.* **66,** 1206–1212.
YAMAGATA, T., SAITO, H., HABUCHI, O., and SUZUKI, S. (1968). Purification and properties of bacterial chondroitinases and chondrosulfatases. *J. Biol. Chem.* **243,** 1523–1535.
ZIMMERMAN, S. B., and SANDEEN, G. (1965). A sensitive assay for pancreatic ribonuclease. *Anal. Biochem.* **10,** 444–449.

AUTHOR INDEX

Numbers in italics indicate the pages on which the complete references are listed.

A

Abbott, J., 37, *39*
Adams, J. M., Jr., 155, 157, *163*
Aksu, O., 53, 54, *57*
Alescio, T., 196, *244*
Ambrose, E. J., 153, *162*
Armstrong, J. L., 14, 17, 28, *41*
Arnold, J. M., 226, *244*
Auerbach, R., 196, *244*
Augusti-Tocco, G., 114, *123*

B

Baker, P. C., 196, 224, 225, 226, 238, *244*
Balinsky, B. I., 15, *39*
Banerjee, S. D., 203, 207, 209, 211, 213, 219, *244*
Barker, S. B., 48, *57*
Barry, J. M., 107, *113*
Barth, L. G., 13, *39*
Barth, L. J., 13, *39*, 59, *99*
Bavetta, L. A., 199, 200, 215, 221, *247*
Becker, R. O., 153, *161*
Benda, P., 114, *123*
Berliner, J., 197, *244*
Bernfield, M. R., 197, 201, 203, 207, 209, 211, 213, 219, 221, *244*
Black, P. H., 221, *249*
Blandau, R. J., 42, *57*
Blumenfeld, O. D., 168, 187, *191*
Boas, N. F., 167, *191*
Boling, J. L., 42, *57*
Bonucci, E., 155, *162*
Borenfreund, E., 167, 172, *192*
Borghese, E., 201, *244*
Bornstein, P., 168, *191*
Brachet, J., 15, *39*
Bringas, P., 199, 200, 215, 221, *247*
Brinster, R. L., 51, *57*
Britten, R. J., 2, *11*, 16, 23, 29, 36, *39*
Brown, D. D., 13, 37, *39*, *40*
Buonassisi, V., 114, *123*, *124*

Buring, K., 125, 126, 127, 156, *162*, *163*
Butter, W. T., 167, *192*
Byers, B., 223, *244*

C

Callan, H. G., 2, *11*
Cameron, J., 199, 200, 215, 221, *247*
Carter, S. B., 226, 227, 228, 233, 236, *244*
Cassini, A., 196, *244*
Cheldelin, V. H., 51, *57*
Chui, D., 199, *247*
Clayton, R. M., 26, *39*
Cloney, R. A., 224, 225, *244*
Coffey, R. G., 51, *57*
Cohen, A. I., 114, *123*
Cohen, J., 170, *192*, 197, 201, 210, 219, 223, 234, *245*
Cohen, J. H., 197, 200, 201, 210, 219, 221, *248*
Coleman, A. W., 60, *99*
Cook, G. M. W., 221, *244*
Coppola, P. T., 46, *57*
Coward, S. J., 27, *39*
Craven, P. L., *163*
Crevasse, L., 47, *57*
Cunningham, L. W., 167, *192*
Cuppy, D., 47, *57*

D

Dameron, F., 196, *244*
Daniel, J. C., 14, 26, 28, *39*, *40*
Danilchenko, A., 167, 180, *192*
Darden, W. H., Jr., 60, 65, 89, 92, 93, *100*
Davidson, E., 170, 173, *193*
Davidson, E. H., 2, *11*
Davis, N. C., 219, *245*
Davison, P. E., 166, *194*
Deffner, G. G., *193*
de la Sierra, J., 127, 146, 155, *163*

AUTHOR INDEX

Delovo, M. C., 203, 204, *246, 247*
DeMeyer, R., 51, *57*
Denis, H., 5, *11*, 16, 27, 31, *39*
Dent, J. N., 115, *124*
DePlaen, J. L., 49, 51, *57*
Dische, L., 167, 170, 172, 173, 174, 177, 179, 180, 181, 183, 184, *192, 194*
Dobell, C., 60, *100*
Dodson, J. W., 170, 182, *192*, 196, 199, 200, *245*
Dorfman, A., 221, *248*
Dorling, J., 204, 205, *247*
Dowell, T. A., 125, 127, 146, 155, 156, 157, 159, *163*
Drake, M. P., *194*
Driscoll, S. G., *112*
Dubuc, F. L., 126, 127, 159, *162, 163*
Dziewiatkowski, D. D., 173, *192*

E

Easty, G. C., 153, *162*
Edds, M. V., 197, 221, *245*
Elias, J. J., 102, *112*
Ellison, M. R., 153, *162*
Entenman, C., 49, *58*
Epler, H., 199, *247*
Ernster, L., 107, *112*
Espinosa de los Monteros, M. A., *112*
Evans, J., 197, 199, 200, 225, *246, 248*

F

Farquhar, M. G., 227, *245*
Farr, A. L., 43, *57*
Fastoe, J. E., 167, *194*
Fell, H. B., 153, *162*
Fessler, J. H., 166, *192*
Fisher, R. A., 47, *57*
Fitton Jackson, S., 165, 169, 190, *192*, 197, *245*
Flickinger, R. A., 12, 13, 14, 15, 16, 17, 26, 27, 28, 30, 31, *39, 40, 41*
Flaxman, B. A., *162*
Freedman, M. L., 30, *39*
Fridhandler, L., 50, 51, *57*
Fukushi, S., 172, 174, 175, *194*, 200, *247*
Furth, J., 115, 116, *124*

G

Gadsden, E. L., 115, *124*
Gallop, P. M., 168, 186, 187, *191, 192*, 210, *247*
Gebhardt, D. O. E., 13, *40*
Gerisch, G., 85, *100*
Gibbins, J. R., 197, *246*
Gillespie, D., 23, *40*
Glock, G. E., 108, *112*
Goldberg, B., 37, *40*, 197, *245*
Goldstein, M., 60, 85, *100*
Golosou, N., 196, *245*
Goodwin, B. C., 153, *162*
Green, H., 37, *40*, 197, *245*
Greene, R. F., 14, 17, 26, 28, 31, *40*
Greengard, O., *112*
Grobstein, C., 153, *162*, 170, *192*, 196, 197, 199, 200, 201, 210, 211, 219, 221, 223, 234, *245, 247*
Gross, J., 37, *40*, 154, *162*, 166, *193*
Gurdon, J. B., 13, *41*
Gustafson, T., 56, *57*

H

Habuchi, O., 206, 213, 221, *249*
Hamburger, V., 27, *40*, 43, *57*
Hamilton, H. L., 43, *57*
Hannig, K., 166, *193*
Harbinson, R. J., 205, *247*
Harper, E., 219, *245*
Harrington, W. F., 165, *193*
Harris, H., 151, *162*
Harrison, R. G., 239, *245*
Hauschka, S. D., 125, *162*, 170, 182, *193*
Hay, E. D., 125, 126, 127, 146, 151, 155, 156, 157, 159, *162, 163*, 169, 178, 189, *193*, 197, *245*
Heidrich, H., 219, *249*
Herrmann, H., 55, 56, *57*
Highberger, J. H., 166, *193*
Hodge, A. J., *193*
Hoffman, P., 170, 173, *193*
Holtfreter, J., 27, *40*, 126, *162*, 196, *245*
Holtzer, H., 37, *39*
Hospelkorn, V. D., 219, *245*
Hsu, T. C., 30, *40*

AUTHOR INDEX

Huggins, C. B., 126, 151, *162*
Hummel, B. C. W., 219, *245*

I

Ito, S., 199, *245*

J

Janet C., 65, 66, 67, 68, 69, *100*
Janners, M. V., 196, *245*
Jones, K. W., 44, *58*
Juergens, W. G., 102, 107, *112*, *113*
Jurist, J. M., Jr., 127, 159, *163*

K

Kafiani, C. A., 30, 35, *40*
Kallman, F., 197, 199, 200, 201, 211, *245*, *246*
Kao, F. T., 115, *124*
Kaplan, D., 170, *193*
Katz, J., 55, *57*
Katzen, R., 49, *58*
Keech, M. G., 166, *193*, *194*
Kefalides, N. A., 171, 172, 173, 175, *193*, 200, *246*
Kelly, D. E., 223, *246*
Kent, P. W., 167, *194*
Kim, U., 116, *124*
Kirk, K. D., 166, *193*
Klein, L., 210, *247*
Klose, J., *40*
Knox, W. E., *112*
Kochert, G., 60, 65, 66, 76, 78, *100*
Koenigsberg, I. R., 125, *162*, 170, 182, *193*
Kofoid, C. A., 85, *100*
Kohl, D. M., 16, 31, *40*
Kohne, D. E., 16, 23, 29, 36, *39*
Kraemer, P. M., 220, *246*
Kratochivil, K., 196, 239, *246*
Krugelis, E., 14, *40*
Kuhn, J., 166, *193*
Kuhn, K., 166, *193*
Kupriyanova, N. S., 30, *40*

L

Lane, J. M., 155, *162*
Lasfargues, E. Y., 102, 107, 108, *113*
Lash, J. W., 153, *162*
Lauth, M. R., 13, 16, 27, 30, *40*
Leader, D. P., 107, *113*

LeBaron, R., 199, 200, 215, 221, *247*
Leblond, C. P., 199, *246*, 247
Lemire, R. J., 53, 54, *57*
Lent, R., 220, *246*
Levine, L., 144, 116, *123*, *124*
Lewis, M. S., 154, *162*
Liedke, K. B., 26, *40*
Lightbody, J., 114, *123*
Lillie, R. D., 203
Linker, A., 170, 173, *193*, 219, *246*
Littna, E., 13, *39*
Loewi, G., *193*
Long, C., 49, *57*
Loring, J. M., 51, *58*
Lowther, D. A., 221, *248*
Lowry, O. H., 43, *57*

M

McBride, O. W., 165, *193*
McCarthy, B. J., 14, 17, 28, 36, *41*
McCracken, M. D., 89, 93, *100*
McIntosh, J. L., 223, *246*
Mackler, B., 53, 54, *57*
McLean, P., 108, *112*
Marks, V., 48, *57*
Martin, A. V. W., 169, *193*
Martin, G. R., 154, *162*, 189, *193*
Masui, Y., 13, *40*
Mathews, M. B., 219, 221, *246*
Mayhew, E., 221, *246*, *248*
Meezan, E., 221, *249*
Meilman, E., 210, *247*
Melnikova, N. L., 35, *40*
Meyer, K., 167, 169, 170, 172, 173, *193*
Miller, F. J., 189, *193*
Mills, E. S., 101, *113*
Miyagi, M., 13, *39*
Morgan, T. H., 10, *11*
Moscona, A. A., 197, *248*
Moser, C. R., 13, *39*
Murray, D. G., 153, *161*

N

Nadol, J. P., Jr., 197, *246*
Nanney, D. L., 153, *162*
Neutra, M., 199, *247*
New, D. A. T., 42, 44, 46, 47, *57*, *58*
Newburgh, R. W., 51, *57*

AUTHOR INDEX

Neyfakh, A. A., 35, *40*
Nicholas, J. S., *40*, 43, *58*
Nieuwkoop, P. D., 13, *40*
Nimni, M. E., 154, *162*
Nogami, H., 125, *162*, *163*

O

Ogi, K., 13, *40*

P

Palade, G. E., 107, *112*, 227, *245*
Palmer, W. M., 50, *57*
Parker, M. L., 114, 116, *124*
Parlebas, J., 183, *194*
Pearce, T. L., 223, 234, *246*
Pfahl, D., 166, *194*
Pier, K. A., 189, *193*
Pierce, G. B., 199, *246*
Piez, K. A., 154, *162*
Pintner, I. J., 60, *100*
Pocock, M. A., 65, 93, *100*
Porter, K. R., 197, 223, *244*, *246*
Powers, J. H., 59, 60, *100*
Powers, M., 189, *193*
Provasoli, L., 60, *100*
Puck, T., 115, *124*

Q

Quintarelli, G., 203, 204, *246*, *247*

R

Rachkus, Y., 30, *40*
Rambourg, A., 199, *246*, *247*
Randall, J. T., 169, *194*
Randall, R. J., 43, *57*
Rappaport, R., 226, *247*
Remington, J. A., *41*
Reoel, J. P., 151, *162*, *163*, 169, 178, 189, *193*, 197, *245*
Ridler, M. A. C., 226, 227, 236, *247*
Rifkind, R. A., 199, *247*
Robbins, P. W., 221, *249*
Robert, B., 183, *194*
Robert, L., 179, 183, 184, *192*, *194*
Robkin, M. A., 46, 47, *58*
Rodén, L., 221, *247*
Rojkind, M., 168, 187, *191*

Rollins, E., 13, *39*
Romanoff, A. L., 55, *58*
Ronzio, R. A., 196, *247*
Rosebrough, N. J., 43, *57*
Rosenberg, J. M., 126, 127, *163*
Rosenthal, M. D., 114, *124*
Rothschild, C., 172, 173, 174, *192*, *194*
Rubin, A. L., 166, *194*
Rudnick, D., 43, *58*
Rutter, W. J., 196, *247*

S

Sahib, M. K., *112*
Saito, H., 206, 213, 221, *249*
Sandeen, G., 221, *249*
Sato, G., 114, 116, 121, *123*, *124*
Saunders, J. W., Jr., 196, *247*
Sayers, E. R., *100*
Schmid, W., 30, *40*
Schmitt, F. O., 166, *193*, *194*
Schroeder, T. E., 196, 224, 226, 238, *244*, *247*
Schuppler, G., 166, *193*
Schwartz, M., 37, *40*
Schwarz, W., 169, 178, 181, *194*
Scott, J. E., 203, 204, 205, *246*, *247*
Searls, R. L., 196, *245*
Seifter, S., 210, 219, 220, *245*, *246*, *247*
Shaw, W. R., *100*
Shepard, T. H., 46, 47, 49, 50, 53, 54, *57*, *58*
Shin, S., 114, *124*
Siekevitz, P., 107, *112*
Silverman, B. F., 126, 127, *163*
Sippel, T. O., 55, *58*
Slavkin, H. C., 199, 200, 215, 221, *247*
Smith, E. L., 219, *245*
Smith, G. L., 226, 227, 236, *247*
Smith, G. M., 61, 96, *100*
Smith, K. D., 14, 17, 28, *41*
Sobel, H., 116, *124*
Speakman, P. T., 166, *194*
Spiegelman, S., 23, *40*
Spiro, R. G., 172, 174, 175, 177, 190, *194*, 200, *247*
Spooner, B. S., 196, 225, 227, 228, 231, 234, 236, 239, *248*, *249*
Spratt, N. T., 55, *58*

Stambrook, P. J., 13, 16, 27, 30, *39*, *40*, *41*
Starr, R. C., 61, 65, 73, 74, 79, 89, 92, 93, 95, 96, 97, *100*
Stein, J. R., 60, *100*
Stein, K. F., 44, 47, *58*
Steinke, J., *112*
Stockdale, F. E., 101, 102, 107, 108, 109, *112*, *113*
Stone, G., 26, *39*
Strates, B. S., 125, 127, 146, 154, 155, 156, 157, 159, *162*, *163*
Stuart, E. S., 197, *248*
Stubblefield, E., 30, *40*
Summerson, W. H., 48, *57*
Suzuki, S., 206, 213, 221, *249*
Sweeney, P. R., 197, 221, *245*
Sweet, W., 114, *123*
Szollosi, D., 226, *248*

T

Taderera, J. V., 196, 239, *248*
Takata, C., 14, *41*
Tamarin, A., 44, *58*
Tanimura, T., 46, 47, 48, 50, *58*
Tashjian, A. H., Jr., 114, 116, 121, *124*
Ten Cate, G., 14, 26, *41*
Thomas, W. C., 154, *162*
Thorp, F. K., 221, *248*
Tilney, L. G., 237, *248*
Timofeeva, M. Y., 30, 35, *40*
Toole, B. P., 221, 243, *248*
Tootle, M. L., 55, 56, *57*
Topper, Y. J., 101, 102, 107, 108, 109, *112*, *113*
Trelstad, R. L., 151, *163*
Trinkaus, J. P., 151, *163*
Turkington, R. W., 121, *124*
Twitty, V. C., 14, *41*

U

Urist, M. R., 125, 126, 127, 146, 151, 154, 155, 156, 157, 159, *162*, *163*

V

Vahouny, G. V., 49, *58*
Vande Berg, W. J., 89, 95, *100*
van de Putte, K. A., 127, *163*
Villee, C. A., 15, *41*, 51, *58*
Vosgian, M. E., *40*

W

Waddington, C. H., 2, *11*, 43, *58*
Wastila, W. B., 50, *57*
Waterman, A. J., 43, *58*
Weber, C. S., *39*
Weiss, L., 221, *246*, *248*
Weiss, P., 125, 145, *163*
Wessells, N. K., 196, 197, 199, 200, 201, 210, 215, 219, 221, 223, 224, 225, 226, 227, 228, 231, 234, 235, 236, 239, *246*, *247*, *248*, *249*
Weston, J. A., 195, *248*
Whiteley, A. H., 36, *41*
Whiteley, H. R., 36, *41*
Wilson, E. B., 1, 10, *11*
Windrum, G. M., 167, *194*
Winzler, R. J., 175, *193*, 200, *246*
Wishnow, R., 114, *124*
Wolff, E., 196, *248*
Wood, G. C., 166, 170, *194*
Wood, H. G., 55, *57*
Woodland, H. R., 13, *41*
Woodlin, A. M., 179, *194*
Wrenn, J. T., 224, 225, 226, 228, 235, *249*
Wu, H. C., 221, *249*
Wünsch, E., 219, *249*

Y

Yamada, K. M., 228, 231, 234, 236, *248*, *249*
Yamada, T., 14, *41*, 153, *163*
Yamagata, T., 206, 213, 221, *249*
Yasumura, Y., 114, 116, 121, *124*
Yates, F., 47, *57*
Ylomans, J. D., 127, *163*
Young, W. C., 42, *57*

Z

Zelmanis, G., 172, *192*
Zelmenis, G., 167, 177, *192*
Zimmerman, S. B., 221, *249*
Zwaan, J., 223, 234, *246*
Zwilling, E., 153, *163*

GLOSSARY OF ABBREVIATIONS

ACTH	adrenal corticotropic hormone
AIB	alpha-amino isobutyric acid
AMP	acid mucopolysaccharide
ATP	adenosine triphosphate
DME	Dulbecco's modified Eagle's Medium
DNA	deoxyribonucleic acid
D-RNA	RNA with base composition similar to DNA, i.e., DNA like RNA
DPNH(NADH)	diphosphopyridine nucleotide
EDTA	ethylenediamine tetraacetate
EGTA	ethelene glycol tetraacetic acid
EM	electron microscopy
ER	endoplasmic reticulum
ETP	electron transport particle
FAD	flavin adenine dinucleotide
FCS	fetal calf serum
FDNB	fluorodinitrobenzene
FMC	fixed mesenchymal cells
FMN	flavin mononucleotide
FSH	follicle stimulating hormone
FUdR	fluorodeoxyuridine
G1	growth-1 phase of the cell division cycle
G2	growth-2 phase of the cell division cycle
G-6-P-D	glucose-6-phosphate dehydrogenase
GNCl	guanidine HCl
GNGl	sialoglylosaminoglycan
I	insulin
MEth(β-MEth)	mercaptoethanol
MPS	mucopolysaccharide
mRNA	messenger ribonucleic acid
NADH(TPN)	nicotinamide adenine dinucleotide
NH	absence of insulin
PAS	periodic acid-Schiff reaction
PTA	phosphotungstic acid
RER	rough endoplasmic reticulum
RNA	ribonucleic acid
RNase	ribonuclease
S-phase	DNA replication (synthesis) phase of the cell cycle
TSH	thyroid stimulating hormone
UV	ultraviolet

SUBJECT INDEX

A

Acellular matrix, see demineralized matrix
Acid mucopolysaccharide (AMP) 172
  acid mucopolysaccharide-protein complexes 213, 243
  at epithelial-mesenchymal junction 203
  at epithelial surface 207, 210
  at morphogenetically active zones 223
  distinguished by hexosamines 203
  distinguished by polyanionic properties 203
  distinguished by susceptibility to β-1-4 hexosaminidases 203
  distinguished by two-step PAS 203
  distinguished by uronic acid residues 203
  histochemical demonstration of 203
  interactions *in vitro* with collagen 221
  in morphogenesis 213, 220
  in morphology 221
  newly syntheized 223
  protein complexes 221
  reaction to PAS 203
Alpha-amino isobutyric acid (AIB)
  effect of insulin 6, 103
  uptake by mammary gland explants 6, 103
Amphibian gastrulation 225
Antecedents to cellular expression, see expression
Antimycin
  inhibiting action of
  Electron Transport System 53
Ascidian metamorphosis
  tadpole-tail retraction in 224
Asexual individuals, see *Volvox*
Asexual reproductive cells, see *Volvox*
Asexual embryo, see *Volvox*
Axon elongation 233, 236

B

Basal lamina
  alteration of 229
  of basement membrane 199
  of cell surface coat 199
  observation by electron microscopy 217
  periodic acid-Schiff (PAS), demonstration of 199
  reappearance of 200
  reconstitution of required for morphogenesis 218
  separation from plasma membrane surface 231
Basement membrane
  absence of in epidermis 200
  basal lamina of 199
  chemical nature of 199
  chemical studies of 199
  collagen fibril assembly in 197
  collagenase action on 197
  determination of by PAS staining 203
  of epithelial cells 7, 199
  at epithelial surface 221
  fibrillar organization in 197
  in morphogenesis 200
  origin of 199
  removal of 200, 211
Biochemical differentiation 1, See also differentiation
Blood circulation
  in rat embryogenesis 43
Bone development 125-163
  bone morphogenesis 126
  determination of 125
Bone induction principle
  for morphogenesis 126
Bone matrix
  crystal formation in 140
  demineralized 127-132, 146
  dentin 145, 153
  glycoproteins in 156
  implantation of 125, 127, 153
  as inductive substratum 126
  lathrytic 149
  in morphogenesis 127
  as morphogenetic substratum 126
  organic 126
  penicillaminated 147

SUBJECT INDEX

as substratum 135, 151
Bone morphogenesis, see also morphogenesis
   embryonic induction of 126
   excavation chamber in 128
   induction of 126
   mitotic activity in 151
   morphogenetic interface during 151
   morphogenetic pattern in 146, 151–154, 156
   morphogenetic substratum for 126
   organic matrix of 126
   substratum acceleration in 126
   substratum for 125, 159
   ultrastructure of substratum and cell interfaces 133
Bone morphogenetic pattern, see bone morphogenesis
Branching pattern
   cytochalasin B and microfilaments 227
   in epithelial glands 8, 195–249
   in morphogenesis 8, 239
   in submaxillary salivary gland of mouse 8, 207
BUdR
Bromodeoxyuridine, BUdR
   in less redudant sequences 37
   for selection of hormone dependency 115
   for self destruction 9, 115

C

Calcification
   apatite microcrystals in 140
   in bone matrix 140, 153, 155, 157, 159
   in collagenous bone matrix 140
   in crystal formation 140
   microglobules in 140
   in morphogenesis 140, 153–159
Capability, embryonic
   detection of 4, 6
      in ontogeny 4, 6, 7
   differentiation of 9
   initiation of 6, 101–113
   maintenance of 9
   prospective 2
   response to 9
   to retain stabilized and differentiated syntheses, selection for 9

Cartilage, see matrix
Cell death, 196
Cell determination, see determination
Cell differentiation, see also differentiation
   in bone morphogenesis 128, 144
   in implants of bone matrix 128
   role of substrata in bone morphogenesis 126
Cell divisions
   number of 6, 12, 14, 16, 27
   rate of 3, 6, 12, 14, 16, 27, 35
   relation to determination 25
   relation to differentiation 13
   separation of rate and number 14
   slower rate of 6, 27
Cell-flattening tendency 8, 227
Cell interaction
   hormones and cell growth 123
Cell lines 114–124
   derivation of 9
   hormone-dependent 9
   hormone-producing 9
Cell populations
   cell migration 233
   shape 225
      3-dimensional changes in 223
   unique, appearance of 7, 195
Cell products
   epigenesis of 2
   initiation of 2
Cell-rounding tendency 8, 207
Cell shape
   asymmetry 223
   cell-surface convolutions 225
   changes in 223
   control during morphogenesis 223
   role of microtubules in 223
Cell-substratum interactions
   cell and bone matrix 125–163
Cell-surface coat 199
Cell and tissue interactions 1, 3
   induction 7, 196
Cellular behavior
   in isolation 5
Cellular changes
   in substrate matrix 10, 144
Cellular commitments
   determination of 5

initiation of 5, 12–41, 74
onset of 5
 cellular and tissue interactions 3
 competence of 3
 determination of 3
 potency of 3
Cellular differentiation 1, See also differentiation
Cellular expression
 artificially amplified 9, 114–124
 dependent upon hormonal stimulation 9, 101–113
 existing, specialized, regulation of 2
 initiation of 2
 regulation of 2
Cellular responsiveness 4
Cellular and transcriptional correlates 5
^{14}C-lactate production
 glucose metabolism 47
^{14}CO$_2$
 glucose metabolism 47
Changing populations of gene copies 5, 17
Changing syntheses 1
 in cell and tissue interactions 11
 cellular and transcriptional correlates in 5
 commitment to 6
 detection of new products and capabilities in ontogeny 4
 dominance of gene copies from nonrepetitive DNA in 8
 in early mammalian embryos 42–58
 effectiveness of gene copies in 6
 of embryo *in vitro* 45
 exponential growth and copies of repetitive DNA in 5
 growth of 11-day, 25-somite rat embryo 45
 initiation of 6
 in initiation of new expressions 6
 intracellular 4
 less redundant gene copies in 5
 maintenance and modulation of previously established expression in 8
 onset of cellular commitments in 5
 oxygen requirement and glucose metabolism, rat embryo 46
 sequelae following well-defined events in 4

Chondrocytes 139
Chondromucoproteins
 acquisition of 10
Chorio-allantoic circulation
 during rat embryogenesis 43
Chromatin
 nuclear 2
 nuclein 1
 program of 2
Chymotrypsin
 effect of on sex inducer in *Volvox* 76
Colchicine
 lens morphogenesis 234
 microtubules, effect on 234
 treatment of epithelium 234
Collagen see also tropocollagen
 adult 168
 alterations of 229
 amino acid composition of 165
 atypical 165
 in bone 141, 154
 in bone matrix 135
 chemical characteristics of 164
 collagenpolypeptides
  shortchain in carbohydrate-rich collagen 183–189
 cross-linking 154, 159
 developmental age of 177
 differences between embryonal and adult types 168
 distribution in morphogenesis 223–227
 embryonic 164–169
 of epithelia 197
  deposition of 201
  in epithelia branching 200
  fibrogenesis 221
  of lung 201
  mesenchyme-epithelial junction 206
  of pancreas 209
  sites for collagen 201
 in extracellular matrix 200
 fibers 135, 140, 221, 223, 241–244
 fibers in basal lamina 199
 interaction with MPS complexes 241, 244
 interfibrillar matrix of 171
 lens capsule 171–177
 in morphogenesis 171, 197, 220
 morphogenetic function of 164–191
 morphology of 164–171

non-calcifying 140
organization 168-178
physical characteristics 164, 168
relation of to carbohydrates 164
skin 166
solubility in bone 155
S-S linkages
of stroma 179-183
in vertebrate eye 164-191
Collagen fibers, see collagen
Collagen fibrils
atypical 165
characteristics of 168
in corneal stroma 178-189
embryonal 168, 171
fibrillation 166
growth of 166
in interfibrillar matrix 166, 169
in morphogenesis 140, 197
native 166
nucleation 166
stromal collagen fibrils organization 187-189
Collagen morphology 164-171
morphogenesis 169-171
role of interfibrillar matrix 169
Collagen synthesis
regulation of by mesenchyme 197
Collagenase 156
collagenase-susceptible materials in morphogenesis 201, 220
cysteine inhibition of 210
epithelia 210
in epithelial morphogenesis 197, 210
in epithelial rudiments 200, 210
high collagenase concentrations 210, 213
for isolation of epithelia 209-214
low collagenase plus hyaluronidase 214
low collagenase treatment 210-218
pancreatic epithelium, effect on 200
purification of 201, 220
sensitivity to 201
for separation of fat and epithelial cells 102
Competence
changes in 27
changes with cell division 28
embryonic, restriction of 38
prolonged 3

Competition hybridization 23
Control mechanisms, see regulation
Corneal stroma 7, 178-189
carbohydrate-rich extractable collagen 179
content and composition of carbohydrate 181
hypothesis about structural significance of extractable collagen 182
complex between sulfated sialoglycosaminoglycans and carbohydrate-rich collagenpolypeptides 183-189
hypothesis about significance of complex for organization of stromal collagen fibrils 182
preparation and analysis 183
presence of shortchain collagenpolypeptides 187
Cytochalasin B
accumulation of multinucleate cells 236
cytokinesis and 227
effect on developing lung 236
effect on filamentous network 234
in epithelial cultures 241
extracellular materials as judged by electron microscopy 229
probe of microfilament systems 227
reversal of 229, 231, 239
in undulating membrane of glia cells 234
Cytochrome C
in DPNH oxidase 53
in electron transport system 52
Cytodifferentiation
appearance of unique cell types 7
organ formation 195
Cytosine arabinoside 28

D

Demineralized bone, see bone matrix
Demineralized bone matrix, see matrix
Demineralized matrix, see matrix
Dentin matrix, see bone matrix, see matrix
Determination
of bone development 125, 151
cell 5
delayed
onset of cellular commitment 3
early, cell 5
inhibition 27

neural 34
in organ development 195
pathways 11
precocious 16
of somatic cells, see also *Volvox* 84
Developmental biologist
definition of 10
Differential cell death 196
Differential rates of division 6
Differentiated cell function
in culture 114
Differentiated syntheses
selection for 9
Differentiation, see also *Volvox*
alteration, pathways of 16
biochemical 1
in bone 126, 128
in bone morphogenesis 125, 128
cellular 1
mesenchyme, in bone 125
neural 13, 34
occurrence of 9
terminal 3
DNA
content of early embryo 13
early replication of
correlated with increase in kinds of DNA synthesized 31
increased transcription from less redundant regions 23
late replicating 12, 30
in amphibian endoderm 34
reduction of transcription 34
nonrepetitive 3, 152, 157
reduced transcription of
during early development of amphibian embryo 30
redundant 12
transcription restricted by late replication 30
DNA-like-RNA
accumulation of
in cytoplasm 27
during cell divisions 29
during development 17, 35
qualitative 17
in adult frog liver 23
changes in

accumulation in cytoplasm 27
conservation in cytoplasm 27
kinds of 5, 16, 22, 27, 36
qualitative 4, 6
quantitative 4, 6
synthesis per cell 27
conservation of 12, 15, 23, 26
conservation in cytoplasm 23
content of
conserved at lower temperatures 28
conserved by slower division rate 29
synthesized at lower temperatures 28
copies
per cell 14
conserved by slower division rate 29
at gastrula and neurula stage 29
synthesized at lower temperature 14, 28
determination of 27
development at later stages 25, 36
epigenesis of 2
exhaustion of 25
in few copies 20
hybridization of 14
kinds of
in control of cellular differentiation 36
fewer, decrease 22, 34
more, increase 27
quantity 22
synthesis of 16
multiple copies of 38
in neurula nuclei 23, 25
nuclear synthesis of 20
populations of 4
quantitative and qualitative changes in 4, 6
redundant 23
short half-life of 15
species of 25
synthesis of 12, 16
changes in 27
control by mitotic age 28
proteins involved in cell maintenance 37
rate and number of cell divisions 13
transcription
from genes arising later in evolution 36
from neurula to larval stage 27
from redundant DNA 23

SUBJECT INDEX 261

by redundant genes 37
transport of
  in cytoplasm 23
  from nucleus to cytoplasm 23
DNA sequences
  copies from 2, 3
  evolutionarily older 38
  less repetitive 3
  more redundant 3
  nonrepetitive 2
  repetitive 3
    evolutionarily old 3
  replication 3
  responsible for cell determination 35
  for sustenance and cell division 5
  transcription of 35
DNA synthesis
  mammary epithelial cells 101, 107
  mammary gland explants from mice 101
  uncoupled from RNA 3
DPNH oxidase
  activity 53
  reoxidation 52
  in somite stage of embryo 56
  terminal electron transport 43

E

Effectiveness of gene copies, see also gene copies
Electron transport system, see terminal electron transport system
Embden-Meyerhof pathway 51, 54
  high rates of anerobic glycolysis 56
Embryo, see also *Volvox*
  asexual 5
  culture method 43
  energy metabolism 42
  heart-rate changes 46
  heart-rate observations 43, 46
  mammalian 42
  oxygen concentrations 46
  oxygen toxicity 46
  protein increment 46
  regions of 6
  whole embryo culture 43
  whole embryo *in vitro* 43–46
  somite stage

anaerobic conditions 55
change in mitochondria number 56
chorio-allantoic circulations 55
DPNH oxidase 56
emergence of oxygen dependence 55
glucose utilization 56
high rates of anerobic glycolysis 56
increase in terminal electron transport system 55
lactate production 56
low activity of Kreb's cycle-electron transport system 56
oxygen requirement 56
pentose phosphate pathway 56
pentose phosphate shunt 55
yolk sac 55
Embryo, *Volvox*
  asexual 68, 71
  female 71
  male 71
  spheroidal 69
Embryogenesis in *Volvox* 61–84
  early 66
  gonidial initials 68
  inversion 69
  somatic initials 68
Embryonal collagen 168
  morphogenetic functions 171
Embryonic determination 27
Embryonic induction
  bone morphogenesis 126
  class of interactions 195
  process of 7
Endoderm cells
  fate of 13
Endoplasmic reticulum, see rough endoplasmic reticulum (RER)
Energy balance 5
Energy metabolism
  in carbohydrate utilization 54
  in mammalian embryos 42
Energy utilization 5
Environmental evocators 1
  influences 10
  manipulations 4
Environmentally induced changes 2
Enzymatic treatments
  Clostridial collagenase preparations 219

degradative activities of enzymes 218–220
effect on intracellular organelles 44
in epithelial morphology 218
Epigenetic changes
cell products 2
Epithelial cells, mammary
responses to insulin 102
Epithelial-mesenchymal interface 199, 203–206
Alcian Blue 7, 203
Epithelial morphogenesis
dependence upon contractile filamentous organelles 239
dependence upon mucopolysaccharide-protein complexes 239
Epithelial morphogenesis, control of 196–227
acid mucopolysaccharide at epithelial-mesenchymal junction 203–207
labeling of epithelial-mesenchymal junction 203–206
staining of epithelial-mesenchymal junction 203–206
Alcian Blue 203
periodic acid-Schiff (PAS) 203
characteristics of surface-associated material 220–223
degradative activities of enzymes 218–220
electron microscopy of epithelia 214–218
extracellular 196–203
intracellular 223–227
Microfilaments and salivary gland morphogenesis 227–239
surface-associated mucopolysaccharide and salivary gland morphogenesis 207–220
mucopolysaccharide at surface of isolated epithelia 210–213
Epithelial morphology
development 196
maintenance 219
Epithelial shape
stabilized 199
Epithelial-mesenchymal interaction
salivary gland 7
Epithelial-mesenchymal interface 7
of embryonic tissues 200
in labeling of epithelial-mesenchymal junction 203–206
in morphogenesis 197
Schiff reactivity 205
Epithelial-mesenchymal junction 199, 203–206
acid mucopolysaccharide 204
Alcian Blue 203
glucosamine 204
staining of 203
uronic acid 204
Ethionine 26
Evocators
environmental, extracellular 1
Exhaustion hybridization, see hybridization
Existing machinery activation 7
Explanted mammary gland, see mammary gland
Exponential growth, see growth
Exponential synthesis, see syntheses
Exported products, see also products
integration of 8
production of 9
Expression
antecedents to 10
biochemical 6
controls for 8
existing 2
initiation of 2, 6
maintenance of 8
modulation of 2
occurrence of 10
previously determined 8
regulation of 2
ultrastructural 6
Extracellular calcifiable matrix, see matrix
Extracellular materials
depletion by low collagenase treatment 217
judged by electron microscopy 229
loss of 240
in morphogenesis 196
as sources of stabilization 235
Extracellular matrix, 151, 199, 215, 221, 241, See also matrix
Extracellular bone matrix, see matrix
Extracellular substrates 199

SUBJECT INDEX

F

Female embryo, see *Volvox*
Fibrillar collagen
  in embryogenesis 7
  interaction with tropocollagen 7
Fibrillar organization
  in morphogenesis 169, 227
Fibrils, see also collagen fibrils
  from tropocollagen 7
Fixed mesenchymal cells (FMC), see also bone matrix, see also mesenchyme cells
  in bone
    dense bodies 134
    electron-dense material (ground substance) 134
    filipodial extensions 134
    filipodial plasmalemma 135
    glycogen bodies 134
    lysomes 134
    mitochondria 134
  in bone morphogenesis 135, 139
Fluorodinitrobenzene (FDNB) 156
Fluorodeoxyuridine (FudR) 32

G

Galactoglucan 7
Gene copies
  changing populations of 5
  continued expression of specific cell properties 9
  effective 9
  effectiveness of 2, 6, 9
  existing, effectiveness of 2, 8
  less redundant
    effectivity of 6
    favoring of 6
  more redundant 3
  multiple
    dominance of 3
    phase out of 6
  old and repetitive
    dominance of 5
  older and more redundant 3
  redundant 5
Gene regulator substances 2
Gene slaves 2

Genes for cell division, sustenance and housekeeping
  evolutionarily old 5
Genome
  active, percent of 16
  activity at early neurula stage 22
Glucorticoids
  hydrocortisone 101
  in mouse mammary epithelial cells 101
  RER 105
Glucose
  appearance of lactate 55
  carbohydrate utilization 54
  conversion to pentose 51
  Embden-Meyerhof pathway 51
  energy metabolism 54
  glycolysis 54
  Kreb's cycle-electron transport system 55
  lactate production 50
  level in serum 48
  metabolic fate 47
  oxygen dependence 55
  pentose phosphate shunt 55
Glucose-6-phosphate dehydrogenase (G-6-P-D)
  cytoplasmic, enzymic activities of 6
  dependence upon insulin 106
Glucose utilization
  comparison of *in vitro* and *in vivo* 52
  in embryo culture 42
  embryo-mother distribution 52
  observations *in vitro* 43
  utilization by rat embryo 49
Glucosamine
  at epitheliomesenchymal junction 206
  incubation of intact glands 213
  radioactivity 214
  at surface of epithelia 211
Glycoprotein
  carbohydrate content 199
  collagen morphology 169
  collagen organization 167
  complexes 183-189
  extraction 179
  high carbohydrate content 199
  large molecular weight 199
  of membranes 221
  polypeptides 183-187

preparation 174
purification 174
Glycosaminoglycans
  developmental age 177
  in morphogenesis of tissue collagen 169
  interfibrillary collagen matrix 169
  separation from sialoglycosaminoglycans 172
Gonidia, see *Volvox*
Gonidial initials, see *Volvox* gonidia
Growth
  exponential 2, 5

H

Hexosaminidases
  morphogenesis 206
Hexosaminoglycans
  in mature collagen 167
Hormone, see also hormone dependency, hormone independency, hormone responsiveness
  exaggerated responsiveness to 9
  stimulation by 4
  withdrawal of 9
Hormone dependency
  Bromodeoxyuridine 117
  cell lines 114–124
  cloned cells 121
  development of hormone-dependent mammary tumor cell line 115
  induction of hormone-dependent tumors 115, 123
  isolation of cell strains 123
  maintenance of differentiated cell function in culture 115
  mammary cells 117
  presence of
    growth hormone 116
    insulin 116
    prolactin 116
  selection for 115, 117
Hormone independency
  cell strains in culture 117
  insulin 116
Hormone responsiveness
  development of cell strains *in vitro* 114–123
  development of hormone dependent mammary tumor cell line 116
  exaggerated 9
  insulin 6
  mammary epithelial cells 116
  selection for hormone dependent cell strains 115
Hyaluronidase
  electron microscopy 214–218
  isolated epithelia 211, 213
  microfilaments 235
  microtubules 235
  salivary gland 206
Hyaluronidase-susceptible material 206
  synthesis by epithelium 201
Hybridization-D-RNA and DNA 14
  exhaustion 14, 20
  to saturation 16

I

Inducer substance, see *Volvox*
Induction, see also cell and tissue interactions
  embryonic, process of 7
Initiating events 4, 6
Insulin
  effect on mammary epithelium 103
  hormone dependent cell strains 102, 106, 123
  hormone independency 116
  insensitivity to 106
  RER resistance to 105
  sensitivity 102–110
  uptake of AIB 103
Insulin responsiveness
  lag period 101, 104
  mammary epithelial cells 103
  mouse gland *in vitro* 110
  RER 106
Interfacial materials
  in morphogenesis 200
Intracellular changes in syntheses 4
Intracellular control of morphogenesis 223–227
Intracellular microfilaments 8
Irrevocable commitments
  of mesenchymal cells 10

SUBJECT INDEX

K

Kreb's citric acid cycle
  electron transport system (ETS) 55
  oxidation method 54
  oxygen requirement 54
  in somite stage of embryo 56

L

Lactate production 50
Lag period 101, 105, 112
Late replications 6
Lens
  lens cup of amphibian embryos 224
  lens tissue shape 234
Lens capsule 171–178
  saccharides as solubilizers of capsular collagen 171
    acid mucopolysaccharide 172
    collagen of 171
    developmental age 177
    disaccharide linked to collagen 172
    sialoglycoproteins 173
      heterogeneity 173
      preparation and purification of a structural glycoprotein fraction 174
      S-S linkages in stabilization of complexes of capsular collagen and glycoproteins 175
LiCl 13
Lung
  cessation of lung morphogenesis 236
  disruption of microfilaments of developing lung 236

M

Macromolecules
  for export 4
Male embryo, see *Volvox*
Mammalian embryos, see also changing metabolism
  in vitro culture 42
Mammary gland
  explanted 4
    epithelial cells of 6
    explants of from mature, virgin, and non-pregnant mouse 101
    induced changes in 8
    from pregnant and non-pregnant mice 101
    organ culture 102
    sensitivity to insulin 101
Mammary tumor cells
  BUDR effects 117
  growth in culture 116
  hormone dependent cell lines 115
Matrix, see also bone matrix
  acellular 9
  cartilage matrix 139
  cellular changes 10
  collagen
    of connective tissues 167
    interfibrillary matrix 169
  demineralized 9
  demineralized bone matrix 132, 149
  dentin 145, 153
  3-dimensional structure in bone 159
  explant, 3-dimensional 9
  exported products of 10
  extracellular bone matrix 151
    disruption 199
    essential material 200
  extracellular, calcifiable 9
  hypertrophic cartilage matrix 154
  implanted substrate 10, 133
  lathyritic bone matrix 149, 154
  morphogenetic 9
  osteoid 154
  penicillaminated bone matrix 154
  resorption of 131
  undenatured 151
Medullary plate
  of amphibian embryos 224
Mesenchymal cells
  amoeboid 9, 125, 128, 133
  explants 125
  transplants 125
Mesenchyme
  blood forming marrow 151
  bone matrix 125
  bone morphogenesis 125
  calcifiable extracellular bone substances 151
  chondromucoproteins 151

colonization of bone matrix 125
development of new populations 126, 133
differentiation in bone 126, 134
migration 125
proliferation 125
in osteogenesis 133
Mesoderm effects
on epithelial cell 238
on epithelial morphogenesis 238
microfilaments 238
organelles 238
Metabolic agents
influence on embryonic determination 27
Metabolic inhibitors 2
Metamorphosis, see ascidian
Microfilaments
assemblies formed by estradiol injection 235
assembly 226
bundles in glandular evaginations 235
in cell morphogenesis 237
class of intracellular organelles 223
contractile rings 239–241
contractility 224
disorganization 227
disruption 240
disruption by cytochalasin 236
function in morphogenesis 225, 235
hormonally induced 226
integrity of microfilament systems 237
local activity 242
localization of 237
loss of 240
in morphogenesis 224
purse-string effect 224
reconstruction 233
in salivary gland epithelial cells 227
synthesis 225
transient appearance of 235
Microglobules
in calcification 140
of electron-dense material 140
of organic material 140
osmiophilic 140
presence of apatite microcrystals 140
Microscopy of epithelia
junctional complex regions 217

junctional complexes 215
results
of collagenase treatment 217
of hyaluronidase treatment 217
Microtubules
in asymmetry cell shape 223
colchicine 234
in developing cells 237
Mineralization
of bone collagen 141
in bone matrix 140
control by osteoblast 141
Mitosis, differential 196
Mitotic age 16
of cells 28
of D-RNA 28
Modulation
controls 8
of existing expressions 2
More to less repetitive DNA sequences, see DNA sequences
Morphogenesis, see also *Volvox*
basal lamina reappearance, prerequisite for 200
basement membrane role 200
bone 125–161
branching 200, 241
cellular basis 226
cessation 221
change in epithelial shapes by intrinsic agents 234
collagen 171, 197, 220
of collagen 164
collagen distribution 201
collagen in 197
content of collagenase-susceptible materials 201
control of cell shape 223
after cytochalasin treatment 244
dependence upon acid mucopolysaccharide-protein complexes 220
effects of collagenase preparation 220
in embryonic organ formation 195
epithelial 196, 209
epithelial branching 197
epithelial control 197
epitheliomesenchymal interactions 196
extracellular control of 196–203

extracellular materials in 196
genome in 126
interfacial materials 200
intra- and extra-cellular bases of 195
loss of 210
mechanism of 223
mucopolysaccharides in 201
normal 241
organelles in 226
organogenesis 196–203
organ-specific 201
promotion of 239
requirement for reconstitution 218
resumption of 239
of salivary gland 223, 227–239
substrata for cell differentiation 126
unique populations of cells 7
volvox 68
Morphogenetic events 4
dictated by certain genes 195
Morphogenetic matrix
3-dimensional 9
explanted 9
extracellular, properties of 220
Morphogenetic mesenchyme control 196
Morphogenetic movement
initiation 225
localization 225
Morphogenetic substratum
in osteogenesis 126
cell-substratum interfaces 144
demineralized bone 125
dentin 145, 153
3-dimensional structure 149
extracted tendon collagen 125
extrinsic substratum 151
kinetics of cell-substratum interactions 150
Mucopolysaccharides
accumulation of 206
attached to collagen molecules 167
of corneal stroma 179
degradation 220
distribution 206, 242
extraction 179–181
labeling at epitheliomesenchymal junction 206
in matrix of connective tissue 167

in morphogenesis 201
mucopolysaccharide precursors 242
mucopolysaccharide-protein removal
from salivary epithelia 241
newly synthesized 206, 242
radioactive 242
revealed by specific staining 242
surface-associated MPS-protein removal 242, 244
at surface of isolated epithelia 210–213
Mucopolysaccharide-protein complex
in epithelial morphogenesis 239
at epithelial surface 239
extracellular deposition of 8
MPS-protein lack in extracellular matrix 241
MPS-protein at morphogenetically quiescent areas 243
MPS-protein removal from salivary epithelia 241
surface-associated MPS-protein 239
Multiple gene copies, see gene copies
Multinucleate cells
accumulation in lung system with presence of cytochalasin 236

N

NADH-Cytochrome C reductase 6, 105
$NaHCO_3$
accelerate mitosis 16
differentiation 13
Nerve cells
axon elongation 236
dissociated from spinal ganglia, spinal cord 236
filamentous material 237
glial cells movement 236
growth cone activity 236
microspike activity 236
network-condensation effect of cytochalasin 237
neurofilaments 236
neurotubules 236
Neutral polysaccharides 7
New expressions
cellular, initiation of 8, 10
initiation of 6, 8, 10

associated with ultrastructural and biochemical changes 6
New phenotypes
  characteristics of 8
  detection of 2
New products 2
  detection of in ontogeny 4
Nuclear chromatin
  program of 2
Nucleation sites 9
Nuclei
  larval 23
  neurala 23
  tailbud 23
Nucleic acid synthesis
  specialization for 5
Nuclein 1, See also Chromatin

O

Ontogenetic events
  in changing syntheses 4
Ontogeny
  capabilities in 4, 6
  new products in 4
Organelles
  contractile 239
  in epithelial morphogenesis 227
  filamentous 239
  in morphogenesis
    hormone-induced assembly of 226
  intracellular
    enzyme treatments of 234
    in epithelial morphogenesis 227
Organogenesis 195
Osteogenesis
  amoeboid mesenchymal cells in 133
  biosynthetic machinery 153
  osteoprogenitor cells 141
  vitamin D-deficiency rickets 146
Osteoid matrix
  collagen cross-linking 154
Osteoprogenitor cells, see also osteogenesis
  alkaline phosphatase in 152
  contact with old bone matrix 141–144
Oviducal epithelium
  of young chicks 226
Oxygen dependence
  emergence in rat embryos 46

Oxygen toxicity
  of rat embryos 46

P

Pancreatic epithelia 197
  acinar differentiation in 200
  collagen accumulation 200
  collagen content 200
  in morphogenesis 200
  sites for collagen 201
  treated with collagenase 200
Parthenogenesis in *Volvox* 92
Pentose phosphate pathway
  glucose metabolism 55
  glucose utilization 50
  importance to embryo 51
  oxygen requirement 54
  C1/C6 ratio in $^{14}CO_2$ production 51
  reduction of nicotinamide adenine dinucleotide phosphate (NADP) 51
  synthesis of ribose precursors of nucleic acids 51
Periodic acid-Schiff (PAS)
  cell-surface coat collagen fiber basal lamina 199
  for evaluating salivary glands 205
  for labeling of epitheliomesenchymal junction 206
  PAS-positive material replacement 199
Periodic acid-Schiff (PAS) staining material
  in basement membrane 200
  on epithelial surface 200
  at epitheliomesenchymal junction 203–207
  interposition in tooth development 200
  on salivary rudiments 203
Phenotypes, cellular 2
  ontogeny of 195
Phosphodiesterase
  effect on sex inducer in *Volvox* 76
Polypeptides, see also collagenpolypeptides
  chains in lens capsule 172
Post-transcriptional controls 10
Products, cellular
  epigenesis 2
  exported 2
  mutation 2

prospective 8
Prolactin
  hormone dependent cell strains 115
Pronase 156
  effect on sex inducer in *Volvox* 76
  separation of gonidia and somatic cells in *Volvox* 66
Prospective capabilities or products 2
Protein carriers 7
Protein content of rat embryos 42, 46

Q

Qualitative accumulation of DNA-like-RNA 17
Qualitative restriction of transcription 16
Quantitative and qualitative changes in DNA-like-RNA 3-6
Quantitative and qualitative differences in RNA synthesis cell-division rate 28

R

Rapid cellular multiplication 2
  exponential growth 3
  gene copies for 9
  prospective potency 3
Rates of cell division 6
  DNA sequences 3
Redundant genes
  activity in D-RNA 37
  fewer 27
  phylogenetically older 37
Regulation
  of cellular expression 2
  control mechanisms for
    cellular determination 34
    transcription and determination 27
  modulation 3
  of specialized cellular expressions 2
Replications, late 6
Reproductive cells, see *Volvox*
Resorption cavity 9
Response to and/or production of special products or capabilities
  maintenance of 9
Responsiveness to hormones 101-113
  exaggerated 9
  insulin 6

mammary epithelial cells 6
Ribonuclease
  effect on sex inducer in *Volvox* 76
RNA, see also DNA-like-RNA
  association with ribosomes and membranes 3
  copies, dominance of 3
  mRNA, masked, inactive 125
  new or unique species of 4
  qualitatively changed 3
  ribosomal 26
RNA synthesis
  precocious 6
  reduced, during S period 34
  uncoupled from DNA 3
  in vitro 16
Rough endoplasmic reticulum (RER) 7
  gland development 101
  insensitivity to insulin 106
  NADH-cytochrome C reductase
    effect of glucocorticoid 105
    effect of insulin 105
  production of secretory and non-secretory proteins 101

S

S and G_1 phases
  of the cell cycle 3, 6, 30
Salivary epithelium stalk
  branching lobules 201
  collagen fibers 221
  interlobular clefts 201
Salivary gland
  electron microscopy 217
  epithelial cells 227
  morphogenesis 207-220, 227-239
  of mouse embryo 196, 201, 210
  staining 204
Salivary mesenchyme 210, 214
Salivary morphogenesis 201, 207-220, 227-239
  in isolated epithelia 210
Salivary rudiments
  stained with PAS 203
Saturation hybridization, see hybridization following well-defined events 4, 101-113
Sequential changes

of competence 12
of determination 12
Sequential syntheses 4, 12–41
Sex differentiation, *Volvox* 78
　initiation of 79
　sexual males 61
Sexual females, *Volvox*
　determination of 5
Sex inducing factor in *Volvox* 61, 95
　assay of inducer 74
　　effects of
　　　carbon dioxide 76–79
　　　chymotrypsin 76
　　　deoxyribonuclease 76
　　　illumination 76–79
　　　phosphodiesterase 76
　　　pronase 76
　　　ribonuclease 76
　　　trypsin 76
　potency of inducer 75
　purification 75
　site of action 81
　specificity 76
Sexual males, see *Volvox*
Sialidase treatment
　of isolated epithelia 213
Sialoglycohexoaminoglycans 7
Sialoglycoproteins 173
Sialoproteins 7
　in bone 156
　in lens capsule 173
Slower division rates
　DNA sequences 3
Somatic cells
　determination in *Volvox* 84
　somatic initials 69
　Volvox 59, 61
Special products in differentiation
　capabilities 7
　expressions 7
　isolation 7
　production of 7
Specialization for exponential synthesis 5
Specialization for nucleic acid synthesis 5
Spheroids, see also *Volvox*
　asexual 5, 61
　female 65
　gonidia 5

male 65
Volvox 61
Spiral cleavage, see *Volvox*
S-S linkages
　in stabilization processes 175–177
　in tropocollagen 165
Stabilized epithelial shape 199
Stabilized syntheses
　selection for 9
Substratum for bone morphogenesis, see also bone morphogenesis
　bone morphogenesis 128
　cell differentiation 128
　cell interfaces and bone morphogenesis 133
　cell-substratum interfaces with cell differentiation 144
　chemical morphology of bone morphogenetic interface 153
　contact between osteoprogenitor cells and old bone matrix 144
　demineralized matrix 146
　hypothetical coupling of collagen cross-linkages and the calcification initiator to produce bone morphogenetic pattern 159
　implants of bone matrix 128
　kinetics of cell-substratum interactions 150
　new bone from normal matrix 145
　ultrastructure of bone morphogenetic interface 157
　ultrastructure of substratum 133
Succinate oxidase
　activity 53
　heart ventricles 55
　succinate oxidation
　　terminal electron transport system 52
Sulfate ester groups
　in acid mucopolysaccharides 203
Synchronized cells
　biochemical experiments 32
Syntheses
　changing 1, 8, 11
　exponential
　　embryonic specialization for 5
　intracellular changes in 4
　nuclear 20

SUBJECT INDEX

ontogenetic events during 4
oscillation of 8
sequential 4
stabilized 9

T

Terminal differentiation 3
Terminal electron transport system 4
  antimycin-sensitive DPNH portion 54
  cytochrome C in ETP 52
  decrease by riboflavin deficiency in rat 54
  DPNH oxidase 42
  electron transport particle (ETP) 52
  flavin adenine dinucleotide (FAD) in ETP 52
  flavin mononucleotide (FMN) in ETP 52
  mitochondria fragmentation 52
  oxidative phosphorylation process 52
  succinic-oxidase system activity 54
Thymidine
  initiation of mammary epithelial cells 107
  rescue 32
  ^{3}H thymidine
    labeling in bone matrix 142
Tissue and cell interactions 1
Tissue sensitivity
  acquisition of 8
Transcription
  controls 10
  correlates 5
  qualitative restriction of 16
  reduction 23
  from redundant larval DNA, amphibian 23
  from redundant sequences 20
Triggering mechanisms 4
Tropocollagen
  influence of sugar polymers on organization of 167
  in growth of collagen fibrils 178
  molecules 7
  organization into native fibrils 167
  S-S bridges 165
  subunit structure 164–191
Trypsin
  effect on sex inducer in *Volvox* 76

Trypsin-pancreatin treatment 199, 201, 209–213, 220
  in morphogenesis 210
  of microfilaments 235
  of microtubules 235

U

Unique cell types
  cytodifferentiation 7
  produced by cellular specialization 195
Unique gene copies
  dominance of 8
Unique populations of cells (morphogenesis) 7
Uronic acid
  at epitheliomesenchymal junction 204
  residues in acid mucopolysaccharides 203

V

Vitamin D-deficiency rickets
  bone morphogenetic property 147
Volvox
  asexual embryo
    development in male and female strains 68
    differentiation 68, 79
    effect of inducer 71
    morphogenesis 68
    reproductive cells 59, 61
  asexual individuals 59–61
  asexual reproduction 59–61
  asexual reproductive cells 61
  deoxyrybonuclease
    effect on inducer 76
  determination of somatic cells 84
  differentiation 79–93
    determination of somatic cells 84
    embryogenesis 61–73
      asexual embryo 68
      early embryogenesis to 32-cell stage 66
      female embryo 71
      male embryo 71
    in embryo 93
    inducer 74–79
      assay 74

effects of carbon dioxide 76
   effects of illumination 76
   potency 75
   purification 75
   specificity 76
  induction of male and female spheroids 93
  induction of males 87
  induction of parthenospores 92
  initiation of sexual response 79–81
   site of action of inducer 81
   spontaneous females 79
  morphogenesis 68
  mutants affecting developmental pattern 81–84
  release of young spheroids 73
  sex 78
  sexual fusion 73
  zygote germination 73
 egg initials 5, 71
 embryogenesis 61–84
  germination 74
  zygote 74
 female embryo
  influence of sexual males' inducer 71
 fertilization 74
 gonidia 5, 68
 gonidial initials in embryos 68
 inducer substance 5, 74–79
 male embryo
  development under influence of sexual males' inducer 71
 parthenogenesis 92
 reproductive cells 59, 61
 sexual females 55–99
 sexual males 55–99
 spheroids
  asexual 61, 80
  development 68
  effect of inducer 65
  female 61
  male 61
  release of 73
  sexual 61
 spiral cleavage pattern 66

W

Well-defined events, see sequelae

Y

Yolk sac circulation, rat embryo
 after explantation 45
 establishment of 46
 sludging of 45